T0137238

Advances in Intelligent Systems and Computing

Volume 894

The series "Advances in Intelligent Systems and Computing" contains publications on theory, applications, and design methods of Intelligent Systems and Intelligent Computing. Virtually all disciplines such as engineering, natural sciences, computer and information science, ICT, economics, business, e-commerce, environment, healthcare, life science are covered. The list of topics spans all the areas of modern intelligent systems and computing such as: computational intelligence, soft computing including neural networks, fuzzy systems, evolutionary computing and the fusion of these paradigms, social intelligence, ambient intelligence, computational neuroscience, artificial life, virtual worlds and society, cognitive science and systems, Perception and Vision, DNA and immune based systems, self-organizing and adaptive systems, e-Learning and teaching, human-centered and human-centric computing, recommender systems, intelligent control, robotics and mechatronics including human-machine teaming, knowledge-based paradigms, learning paradigms, machine ethics, intelligent data analysis, knowledge management, intelligent agents, intelligent decision making and support, intelligent network security, trust management, interactive entertainment, Web intelligence and multimedia.

The publications within "Advances in Intelligent Systems and Computing" are primarily proceedings of important conferences, symposia and congresses. They cover significant recent developments in the field, both of a foundational and applicable character. An important characteristic feature of the series is the short publication time and world-wide distribution. This permits a rapid and broad dissemination of research results.

**** Indexing: The books of this series are submitted to ISI Proceedings, EI-Compendex, DBLP, SCOPUS, Google Scholar and Springerlink ****

More information about this series at http://www.springer.com/series/11156

Joan Carles Ferrer-Comalat ·
Salvador Linares-Mustarós ·
José M. Merigó · Janusz Kacprzyk
Editors

Modelling and Simulation in Management Sciences

Proceedings of the International Conference on Modelling and Simulation in Management Sciences (MS-18)

Editors
Joan Carles Ferrer-Comalat
Department of Business Administration
University of Girona
Girona, Spain

Salvador Linares-Mustarós
Department of Business Administration
University of Girona
Girona, Spain

José M. Merigó
Department of Management Control
and Information Systems
University of Chile
Santiago, Chile

School of Information, Systems
and Modelling
University of Technology Sydney
Sydney, Australia

Janusz Kacprzyk
Systems Research Institute
Polish Academy of Sciences
Warsaw, Poland

ISSN 2194-5357 ISSN 2194-5365 (electronic)
Advances in Intelligent Systems and Computing
ISBN 978-3-030-15412-7 ISBN 978-3-030-15413-4 (eBook)
https://doi.org/10.1007/978-3-030-15413-4

Library of Congress Control Number: 2019933728

This Springer imprint is published by the registered company Springer Nature Switzerland AG
The registered company address is: Gewerbestrasse 11, 6330 Cham, Switzerland

Preface

The Association for the Advancement of Modelling and Simulation Techniques in Enterprises (AMSE) and University of Girona are pleased to present the main results of the International Conference of Modelling and Simulation in Engineering, Economics, and Management, held in Girona, Spain, 28 to 29 June 2018, through this book of proceedings published with Springer in the Series: Advances in Intelligent Systems and Computing.

In this edition of the MS International Conference, we gave special attention to the use of bibliometrics in social science.

Sixteen papers from 17 universities around the world constitute the MS 2018 Girona proceedings. The book mainly presents papers with a strong orientation to modelling and simulation in general fields of research. But it also covers some papers close to the field of bibliometrics in social sciences.

We would like to thank all the contributors, referees and the scientific and honorary committees for their kind cooperation with MS'18 Girona. Finally, we would like to express our gratitude to the members of Springer for their support in publishing the book.

November 2018

Joan Carles Ferrer-Comalat
Salvador Linares-Mustarós
José M. Merigó
Janusz Kacprzyk

MS'2018 Girona Congress

Modelling and Simulation in Management Sciences 28–29, June 2018, GIRONA (Spain)

Honorary Committee

Jaume Gil Aluja	President of AMSE and President of the Royal Academy of Economic and Financial Sciences, Spain
Joaquim Salvi Mas	Chancellor at the University of Girona, Spain
Anna Garriga Ripoll	Dean in the Faculty of Business and Economics at the University of Girona, Spain

Co-chairs

Xavier Bertran Roura	University of Girona, Spain
Joan Carles Ferrer Comalat	University of Girona, Spain
Anna M. Gil Lafuente	University of Barcelona, Spain
Salvador Linares Mustarós	University of Girona, Spain
José M. Merigó Lindahl	University of Chile, Chile and University of Technology Sydney, Australia

Organizing Committee

Francisco Javier Arroyo	University of Barcelona, Spain
Christian Berger-Vachon	University of Lyon, France
Fabio Blanco	Pedagogical and Technological University of Colombia, Colombia

Elvira Cassú Serra	University of Girona, Spain
Marc Carreras Pijuan	University of Girona, Spain
Dolors Corominas Coll	University of Girona, Spain
M. Àngels Farreras Noguer	University of Girona, Spain
Jaime Gil Lafuente	University of Barcelona, Spain
Carlos Lopez Avellaneda	University of Girona, Spain
Xavier Molas Colomer	University of Girona, Spain
Elena Rondós Casas	University of Girona, Spain
Josep Viñas Xifra	University of Girona, Spain
Emili Vizuete	University of Barcelona, Spain
Binyamin Yusoff	University of Barcelona, Spain

The AMSE 2018 Conference is supported by:

Contents

Intellectual Capital Management and the Innovation Process: Does One Size Fit All?

Juan Carlos Salazar-Elena[1], Christian A. Cancino[2(✉)], Asunción López López[1], and José Guimón de Ros[1]

[1] Department of Economic Structure and Development Economics, Autonomous University of Madrid, Campus de Cantoblanco, Madrid, Spain
{juancarlos.salazar,asuncion.lopez.lopez, jose.guimon}@uam.es

[2] Department of Management Control and Information Systems, University of Chile, Diagonal Paraguay 257, Santiago, Chile
cancino@fen.uchile.cl

Abstract. Intellectual Capital Management (ICM) research has focused predominantly on demonstrating the link between ICM and innovative performance from a general, context-free perspective. This paper analyses deeper implications of ICM showing the inadequacy of this one-size-fits-all approach, and introduces a more realistic view of the role of ICM on innovation process. We propose an analytical framework where differences in ICM strategies are explained by differences in the environment faced by firms and characterized by disparate levels of threats posed by potential imitators. Although legal protection of intellectual property rights has been traditionally the center of attention on this topic, the firm's strategy might be far more complex involving alternative ICM activities guided by the incentives of the environment. Using panel data for Spanish SMEs over the 7-year period between 2008 and 2014, we provide evidence consistent with this approach. Empirical analysis shows that certain ICM strategies are profitable in specific environments, but might be very unattractive in other contexts where other strategies prove to be more efficient. These findings contribute to the development of a more nuanced interpretation of the role played by ICM on the innovation process.

Keywords: Intellectual Capital Management · R&D · appropriability regime · Innovation process

1 Introduction

Although there is a broad consensus about the importance of Intellectual Capital Management (ICM) as a source of value creation in today's economies, recent research suggests that the literature has focused predominantly on demonstrating the link between ICM and innovative performance from a general, context-free perspective (Mouritsen 2006; Guthrie et al. 2012; Dumay and Garanina 2013). As these authors state, this perspective is rooted on the mistaken belief that ICM will yield necessarily greater profits, regardless of the context where the empirical analysis takes place.

J. C. Ferrer-Comalat et al. (Eds.): MS-18 2018, AISC 894, pp. 1–12, 2020.
https://doi.org/10.1007/978-3-030-15413-4_1

This paper analyses deeper implications of ICM showing the inadequacy of this one-size-fits-all approach, and introduces a more realistic view of the role played by ICM on innovation processes. We propose an analytical framework where the differences in ICM strategies followed by firms in order to profit from innovative activities are explained by differences in the environments they face and characterized by disparate levels of threats posed by potential imitators. Although legal protection of intellectual property rights has been traditionally the center of attention on this topic, the firm's strategy might be far more complex involving alternative ICM activities guided by the incentives of the environment. Under this approach some profitable ICM strategies in specific environments (e.g. patenting) might be, in contrast, very unattractive in other contexts where other strategies prove to be significantly more efficient (e.g. developing productive capabilities or market linkages). It is in this sense that we offer a context-dependent analysis of ICM rather than the one-size-fits-all approach prevalent in the literature.

Schumpeter (1950) and Arrow (1962) introduced the "appropriability problem" in the theoretical discussion of innovation, explaining that certain degree of monopoly might be desirable to foster innovation. This theoretical problem has been typically used to justify the introduction of intellectual property rights to guarantee the appropriation of benefits from innovations. However, it is evident that legal protection of intellectual property rights is not a perfect strategy in certain contexts. Teece's (1986) seminal work, Profiting from technological innovation, showed that the appropriation strategy of innovators might be quite complex, involving managerial decisions conditioned by the environment. In particular, innovators' strategy must be shaped by what he called the "appropriability regime". This regime refers to the environmental factors, different from firm and market structure, that govern an innovator's ability to capture the profits generated by an innovation. According to Teece (1986), the most important dimensions of such a regime, are the nature of the technology and the efficacy of legal mechanisms of protection.

In this paper we integrate Teece's approach with the ICM literature to develop a non-trivial explanation of the role of ICM strategy in firms' innovation plans. In order to analyze typical responses of firms -in terms of ICM strategies- related to specific environments, we follow the sectorial taxonomy proposed by Castellacci (2008), that introduces differences in appropriability regimes, highlighting the relevance of different forms of intellectual property protection (patents, trademarks, or copyright) within each sector. Two sectors of activity, intimately related but with significant differences in their appropriability regimes, are used in our empirical analysis: hardware and software industries. The former relies significantly more on intellectual property protection than the later. Using panel data for Spanish SMEs over the 7-year period between 2008 and 2014, we show that certain ICM strategies are profitable in specific environments, but might be unattractive in other contexts where other strategies prevail. These findings contribute to the development of a more precise description of the role of ICM on the innovation process.

This paper is organized as follows. In Sect. 2 we present a critique of the current state of ICM research is discussed. Section 3 presents the use of Teece's approach to explain the differentiated role of ICM, affecting innovation process not only through its role on the "invention plan", but also through the strategy followed by the firm to effectively appropriate the benefits generated by those inventions. Section 4 discusses the appropriability regime concept, emphasizing the way it might induce firms to adopt different ICM strategies. Section 5 explains research design and methods. Section 6 presents the results of the empirical analysis. Finally, Sect. 7 provides a discussion of the managerial and policy implications of the results.

2 Intellectual Capital and Innovation: Overcoming Context-Free Theorizing

The concept of intellectual capital (IC), or intangible assets, is relatively new in the economic literature. It refers to those non-physical assets with three core characteristics: they are a source of probable future economic benefits, have no physical embodiment and, to some extent, may be retained and managed by companies (OCDE 2011).

Although in the past the concept of intangible asset was primarily related to R&D and intellectual property rights (such as patents and trademarks), the extension of innovative activities to other areas beyond the purely technological ones has led to an expansion of this concept. Nowadays it is widely accepted that IC consists of three interrelated bodies of knowledge: capabilities and skills of the members of the team; structured/codified knowledge owned by the firm (such as production processes, internal procedures, results of R&D activities, or intellectual property protection); and the set of relations established with other agents or organizations outside the firm. These three bodies of knowledge are usually called human capital, structural (or organizational) capital and relational capital, respectively (Cañibano et al. 2002).

The specialized literature has highlighted the effect of IC on the innovation process. Some authors have emphasized that organizations that develop and exploit effectively their IC have a competitive advantage (Steward 1997). Becerra et al. (2008) state that an effective ICM can avoid unauthorized knowledge transfers, which is one of the major risks faced by innovative firms. Along the same lines, other recent studies have found a positive relationship between IC management and innovation (Henry 2013; Kremp and Mairesse 2004; Mangiarotti 2012). Marvel and Lumpkin (2007) studied the role of experience, education and prior knowledge on innovation outcomes, and De Winne and Sels (2010) show that human capital (of managers and employees) and human resource management are important determinants of innovation in start-ups. There is a vast literature on the effect of R&D activities and patents on innovation performance, but we can also find studies on the effect of other kind of structured knowledge owned by the firm. As Huchzermeier and Loch (2001) pointed out, there are different sources of uncertainty in R&D activities (market payoffs, project budgets, product performance, market requirements, and project schedules), and management team's ability to adapt processes and procedures is key to improve risk management in R&D projects. Finally, there are several studies analyzing the relative impact of R&D collaborations and participation in networks on innovation performance (Ahuja 2000; Reagans and Zuckerman 2001; Belderbos et al. 2004; Czarnitzki et al. 2007; Un et al. 2010).

The existing literature on intangibles has been primarily devoted to emphasize the impact of IC on firm performance in an ostensive/general way and also to design new methods to measure IC. These two main concerns of research practices corresponds to what Guthrie et al. (2012) denominate, respectively, the first and second stage of research on IC, focused on "revealing" the importance of intellectual capital to create competitive advantages, and on the design of indicators to measure and report IC within the firm. In both stages, the empirical evidence is insufficient or inconclusive, although there is a broad consensus on the importance of intangible assets as a source of value creation (Dumay and Garanina 2013). As discussed in Dumay (2012), IC research has not reached the point where it can be stated that managing IC leads to greater profitability because of the inability to make causal links between IC and value creation.

This paper is ascribed to what Guthrie et al. (2012) calls the "third stage" of IC research. This new stage, just in its infancy, attempts to provide deeper managerial implications, avoiding the general and somehow tautological perspective that simply postulates that ICM boosts innovative performance across the board. We sustain that ICM is part of the business strategy to profit from innovative activities and this strategy is clearly conditioned by the firm's context. Our analysis focuses on differentiating between IC strategies of innovative firms in several contexts and rationalizing them as differences in business strategies to appropriate the benefits of their innovation activity. In this sense, our contribution to the literature is to unpack previous discussion on links between ICM and innovation, analyzing the role of ICM in a context-dependent strategy. This enables us to transcend the simple argument that "the more knowledge the better" and introduce the analysis of ICM into high-level debates about profitability of specific IC strategies, organization and market structure, and public policy.

3 Intellectual Capital, What for?

As stated before, much of the specialized literature has focused on demonstrating the link between ICM and innovative performance in a very general, context-free perspective. In fact, context-free theorizing has been very popular in management and accounting research (Llewelyn 2003). However, little attention has been paid to the specific role played by each element of ICM in the strategy of the innovative firm.

We follow the approach of Teece (1986), assuming that innovative firms are solving a problem: How to profit from their innovations? The contribution of Teece (1986) has to do with the different strategies that a firm can choose to guarantee the appropriation of the rents generated by innovations, conditioned by the characteristics of the knowledge embedded in the potential innovation (tacitness, complexity, observability), the legal instruments to protect innovations (patents, copyrights, designs, etc.), and the complementary activities or capabilities available to the innovator (such as marketing, competitive manufacturing, after-sales support, etc.). We argue that in order to understand the role of ICM in innovative activities, we need to be able to see how it fits in this different «appropriation strategies».

Elaborating on the argument of Teece (1986), there are at least two types of activities needed to profit from innovations:

- Inventive activities, to produce (or acquire) the "core knowledge" embedded in a potential innovation
- Appropriation activities, to profit from the creation of this core knowledge

It is obvious that not all firms engage in inventive activities (e.g. not all firms try to develop new products). And we've learnt from Teece that a successful innovation cannot be assured if the firms focus solely on inventive activities, given the need for managing and developing complementary assets in order to profit from innovation activities.

We propose to classify the role of the different ICM activities regarding their relation with these two types of activities. While certain ICM activities will be directly related with the innovation process, through their role in shaping and executing activities specifically aimed to the development of the core knowledge embedded in potential innovations (as context-free theorizing research has tried to prove), other activities might have a more complex, context-dependent relation with innovation process, e.g. providing services that might improve the ability of the innovative firm to appropriate the profits from its innovations in different specific contexts.

There are very good reasons to expect that ICM might be context-dependent, since the threats and opportunities for innovative firms might not be the same everywhere and the strategy of the firm must discriminate among ICM activities. Take, for instance, the case of tacit knowledge flowing with the mobility of workers. In this case the management of human capital in the company might be essential to guarantee the appropriation of profits from R&D (Hurmelinna-Kaukkanen and Puumalainen 2007; Hurmelinna-Laukkanen et al. 2007; Casper and Whitley 2004). Relational capital might also affect the ability of firms to appropriate the rents of innovation. As Henttonen et al. (2016) observe, collaboration with other partners for R&D activities might generate risks of spreading knowledge and misappropriation of the generated value. This risk can be managed by the firm using different strategies, alongside the mechanisms of protection of intellectual property: lead-time innovation, continuous incremental innovations, cost reduction of innovations, etc. (Pérez-Cano 2013).

We consider that ICM activities must coordinate the firm strategy to profit from the creation of new knowledge, depending on the context of the firm. For example, in an "ideal" (but not necessarily desirable) context where legal framework offers perfect intellectual property protection, inventive activities might be sufficient for the innovative firm, since the new knowledge can be sold or licensed. In this sense, this innovative firm might not need to develop manufacturing capabilities, market linkages, or other type of complementary assets to profit from innovations. As we move away from this ideal case, the threat posed by imitators forces the innovative firm to develop more complex strategies. We assume that ICM strategy is a mixed combination of the following activities:

- Management of the development of the «core knowledge» embedded in a potential new product invention
- Intellectual property protection
- Management of production processes and organizational procedures
- Management of commercialization

While the first one is clearly an Inventive activity, the three remaining activities are what we denominate Appropriation activities. The attractiveness of each possible combination of these four activities will depend on the context faced by the firm, and on the nature of the innovation.

4 ICM and Appropriability Regimes

It was Schumpeter (1950) and Arrow (1962) who introduced the "appropriability problem" in the theoretical discussion of innovation, linking the profitability of innovation with the market structure, in the sense that certain degree of monopoly might be desirable to foster innovation. This theoretical problem was conceptually tackled with the introduction of an ideal patent guaranteeing the appropriation of the benefits generated by innovations. However, as it is well known, legal protection of intellectual property fails to be efficient in certain contexts. As Teece (1986) argues, the appropriation strategy of innovators is far more complex, involving important managerial decisions conditioned by the environment.

The strategy to successfully commercialize an innovation will be strongly conditioned by the environment of the firm. Teece (1986) illustrated how the concept of "appropriability regimes" is useful to understand the different types of environments or contexts faced by innovative firms.

The appropriability regime is a theoretical construction assessing the threats for innovative firms posed by potential imitators. This regime can move within a range that goes from "tight" to "weak", where the former indicates a regime where imitation is difficult (because of the legal framework or the complexity of the knowledge involved in the innovation), while the latter is a situation where it is easy for competitors to copy innovations.

Castellacci (2008) proposed a sectorial taxonomy that highlights the relevance of different forms of intellectual property protection (patents, trademarks, or copyright) within each sector. For example, sectors like Electronics or Machinery rely substantially more on patents than sectors like Software or Engineering. These differences induce different needs from the perspective of ICM: where intellectual property protection is not a profitable strategy (as might be the case of these latter sectors of activity) other types of ICM must emerge to accomplish the goals of innovative firms. In Fig. 1, we draw this possibility of differentiated profitability of ICM strategies as a function of the appropriability regime.

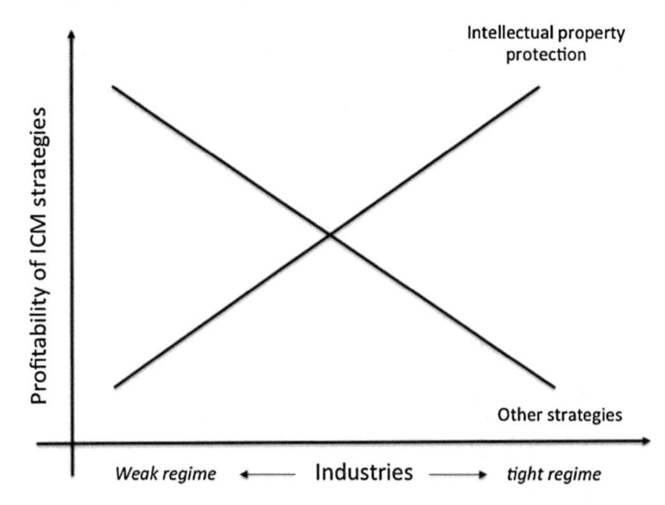

Fig. 1. Differentiated profitability of ICM strategies. Source: Own elaboration

5 Questions and Research Design

In this paper we posit that ICM activities cannot be studied through context-free theorizing, since this type of activities have an organic relation with the specific context faced by the firm. In this sense, and contrary to the tradition of literature relating ICM and innovation, we propose that not all ICM activities are necessarily desirable in all contexts. To challenge this view we ask: does the appropriability regime faced by the firm influence the impact of ICM on innovation activities? And, if this is the case, is it possible to find clear patterns for ICM responses from companies, which can be explained by differences in appropriability regimes?

We propose a semi-log model, where the dependent variable, called RDown, is the natural logarithm of R&D expenditure financed with the own funds of the firm (i.e., excluding funds from other private or public organizations). This variable will be explained by other variables divided in two groups: inventive activities and appropriation activities. The explanation provided by appropriation activities (divided in three subgroups: productive capabilities, commercialization and intellectual property) shows the non-trivial relation of the activities to appropriate profits with R&D effort. All variables are presented in Table 1. Table 2 shows the averages of the variables used in each industry analyzed.

Table 1. Variables of the model. Source: Own elaboration

Variable	Type	Description
Rdown	Dependent variable	In-house R&D expenditure financed with own funds (natural logarithm)
ln_size	Control variable	Num. Of employees (natural logarithm)
innprod	Innovation activities	Product innovation in the period from (t – 2) to t
innproc	Innovation activities	Process innovation in the period from (t – 2) to t
pat	Intellectual property protection	Patent applications
usoautor	Intellectual property protection	Use of copyrights
ln_prod	Productive capabilities	Labour productivity (natural logarithm)
market	Commercialization activities	Activities of market introduction of innovations
inn_comer	Commercialization activities	Marketing innovations
coopcli	Commercialization activities	Cooperation for innovation with costumers

The strategy of this paper to answer our research questions is to test the relation of appropriation activities with R&D profitability under different appropriability regimes, following the sectorial taxonomy proposed in Castellacci (2008). In order to simplify the exposition of the results, we shall focus on two sectors of activity: Software and Hardware. The first is classified, as having a weak appropriability regime where patent protection is not a typical strategy, as opposed to the second where patenting appears to be a profitable strategy. In this sense, our study might be seen as a case study, where empirical evidence is intended to show the plausibility of the relation between the environment of the firm and the IC strategy. The justification for the choice of these two sectors is that they appear to be based on very different strategies regarding intellectual property protection; therefore we expect to find not only confirmation in these two sectors (i.e., showing that patenting is a more profitable strategy in Hardware than in Software industry), but also to prove that this difference induces another strategy in Software.

To test these relations, we use of the Spanish Panel of Technological Innovation (PITEC) –a CIS-type survey–, selecting a sample of Spanish SMEs over the 7-year period between 2008 and 2014. Given the fact that the aim of the study is to analyze the role of certain inputs in the profitability of R&D devoted to the generation of new products, the sample only includes firms declaring innovative activities in at least three years during the period considered. In order to solve potential endogeneity problems with the variables included in the model. Given the nature of our dependent variable, the models are estimated using Tobit model for panel data.

6 Results

Table 2 presents the main results of the estimated model for each sector. We have to underline that this table is interesting not only because it identifies the variables that best explain the behavior of the dependent variable, but also because those that are not relevant reflect differences in the use of mechanisms of appropriation of the rents of R&D activities. Using the information from Table 2, we can state that patenting is in fact a profitable strategy (more in the case of Hardware, as predicted by theory), but the successful strategy of firms from Software industry include other type of activities such as the use of copyright, productive capabilities or commercialization activities.

In terms of IC, our results reflect, first the existence of a positive relationship between many of the variables of IC and R&D, because of the positive effect of these elements on the ability to appropriate the returns from R&D activities. Second, that companies develop different strategies to benefit from their R&D, depending on their specific circumstances. For the hardware sector, the use of instruments of protection of intellectual property such as patents, has a positive and significant impact on the performance of R&D, while in the case of software companies it is also the availability of complementary production and marketing assets that increases the capacity of appropriation.

Table 2. Results of the estimated model

Explanatory variables	Hardware		Software	
	Coef.	P > t	Coef.	P > t
ln_size	4.596	0.000	2.510	0.000
ln_size2	–0.431	0.000	–0.218	0.000
innprod	1.430	0.000	1.469	0.000
innproc	0.509	0.036	0.289	0.156
pat	1.291	0.000	1.099	0.004
usoautor	0.355	0.685	1.986	0.001
ln_prod	0.235	0.210	0.480	0.002
market	0.285	0.171	0.595	0.003
inn_comer	0.424	0.059	1.184	0.000
coopcli	0.298	0.307	1.052	0.000
Year				
2009	–0.490	0.106	–0.349	0.218
2010	–0.602	0.047	–0.744	0.009
2011	–0.824	0.007	–0.539	0.057
2012	–0.814	0.010	–0.334	0.245
2013	–0.873	0.006	–0.736	0.012
2014	–1.376	0.000	–1.412	0.000
Cons.	–3.498	0.148	–3.466	0.070
Num. of obs.	1,428		2,869	
Num. of firms	263		494	
R-squared	0.275		0.152	

A summary of these results is presented schematically in Fig. 2, replicating the idea expressed in Fig. 1.

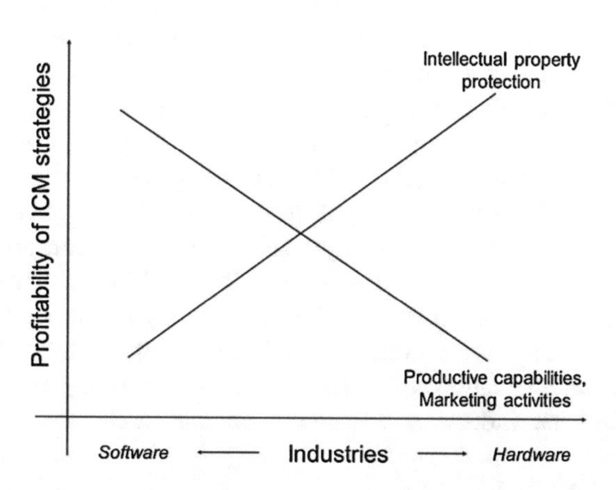

Fig. 2. Profitability of ICM strategies. Source: Own elaboration

7 Conclusions

The aim of this research was to answer the following two questions: Does the appropriability regime faced by the firm influence the impact of ICM on innovation activities? And, if this is the case, is it possible to find clear patterns for ICM responses from companies that can be explained by differences in appropriability regimes? Regarding the first question, evidence is consistent with the fact that there is a differentiated impact of ICM activities across sectors. In the Hardware sector, where patenting is profitable, in terms of R&D the size and marketing activities are not as relevant as in the case of the Software industry.

Indeed, the analysis has shown that, as noted in the taxonomy proposed by Castellacci (2008), appropriability is an element of sectorial differentiation in the sense that different economic activities involve different strategies on legal protection of IC. Since the elements of IC include instruments such as patents, licenses, etc. it can be argued that the capital structure between different sectors is due to the way companies try to appropriate the rents of innovation.

Regarding the second question, our empirical results might be rationalized through the existence of an appropriation strategy whereby firms try to profit from their R&D activities making use of a more diverse ICM beyond the intellectual property protection, either developing bigger productive capabilities or stronger connections with the market.

Our analysis also reveals that companies follow different strategies to pursue the benefits from innovation, even though there are sectors that have a greater supply of complementary assets. We have shown that the hardware sector, for which the appropriability regime is stronger, has higher levels of complementary assets both in terms of production and marketing, but the strategic importance of IC management with respect to the software sector is significantly lower.

This paper has contributed to improving our understanding of the role of IC, and its different elements, on innovative activity. The results confirm that IC management generates positive results through its positive effect on appropriation capabilities. Moreover, this work has contributed towards the development of the "third phase" of IC literature (Guthrie et al. 2012), which seeks to explain the causal relationship between IC and strategic management.

From a public policy perspective, contemporary thinking on innovation policy tends to focus on the generic promotion of the creation of the core knowledge embedded in potential new products, and on the strengthening of intellectual property protection. But patents (and other legal instruments), while offering considerable protection to some products, do not guarantee complete appropriability of some innovations especially in certain sectors. And, as Teece (1986) showed, profiting from innovations may depend not only on the legal protection offered by intellectual property rights, but also on the availability of other complementary assets and capabilities. In this sense, innovation policy becomes more closely aligned with the strategic analysis of markets and industries, considering alternative choices for public policy.

References

Ahuja, G.: Collaboration networks, structural holes, and innovation: a longitudinal study. Adm. Sci. Q. **45**(3), 425–455 (2000)

Arrow, K.J.: Economic welfare and the allocation of resources for invention. In: Nelson, R. (ed.) The Rate and Direction of Inventive Activity, pp. 609–625. Princeton University Press, Princeton (1962)

Becerra, M., Lunnan, R., Huemer, L.: Trustworthiness, risk and transfer of tacit and explicit knowledge between alliance partners. J. Manage. Stud. **45**(4), 691–713 (2008)

Belderbos, R., Carree, M., Lokshin, B.: Cooperative R&D and firm performance. Res. Policy **33**(10), 1477–1492 (2004)

Cañibano, L., Sánchez, P., Sánchez-Ayuso, M., Chaminade, C.: Guidelines for Managing and Reporting on Intangibles. Vodafone Foundation, Madrid (2002)

Casper, S., Whitley, R.: Managing competences in entrepreneurial technology firms: a comparative institutional analysis of Germany, Sweden and the UK. Res. Policy **33**, 89–106 (2004)

Castellacci, F.: Technological paradigms, regimes and trajectories: manufacturing and service industries in a new taxonomy of sectoral patterns of innovation. Res. Policy **37**, 978–994 (2008)

Czarnitzki, D., Ebersberger, B., Fier, A.: The relationship between R&D collaboration, subsidies and R&D performance: empirical evidence from Finland and Germany. J. Appl. Econometrics **22**(7), 1347–1366 (2007)

De Winne, S., Sels, L.: Interrelationships between human capital, HRM and innovation in Belgian start-ups aiming at an innovation strategy. Int. J. Hum. Resour. Manage. **21**(11), 1863–1883 (2010)

Dumay, J.: Grand theories as barriers to using IC concepts. J. Intellect. Capital **13**(1), 4–15 (2012)

Dumay, J., Garanina, T.: Intellectual capital research: a critical examination of the third stage. J. Intellect Capital **14**(1), 10–25 (2013)

Guthrie, J., Ricceri, F., Dumay, J.: Reflections and projections: a decade of intellectual capital accounting research. Br. Account. Rev. **44**(2), 70 (2012)

Henry, L.: Intellectual capital in a recession: evidence from IK SMEs. J. Intellect. Capital **14**(1), 84–104 (2013)

Henttonen, K., Hurmelinna-Laukkanen, P., Ritala, P.: Managing the appropriability of R&D collaboration. R&D Manage. **46**(S1), 145–158 (2016)

Huchzermeier, A., Loch, C.H.: Project management under risk: using the real options approach to evaluate flexibility in R&D. Manage. Sci. **47**(1), 85–101 (2001)

Hurmelinna-Kaukkanen, P., Puumalainen, K.: Nature and dynamics of appropriability: strategies for appropriating returns on innovation. R&D Manage. **37**(2), 95–112 (2007)

Hurmelinna-Laukkanen, P., Kyláheiko, K., Jauhiainen, T.: The Janus face of the appropriability regime in the protection of innovations: theoretical re-appraisal and empirical analysis. Technovation **27**, 133–144 (2007)

Kremp, E., Mairesse, J.: Knowledge management, innovation and productivity: a firm level exploration based on French manufacturing CIS3 data. NBER Working Paper Series, n° 10237 (2004)

Llewelyn, S.: What counts as 'theory' in qualitative management and accounting research? Introducing five levels of theorizing. J. Acc. Auditing Accountability **16**(4), 662–708 (2003)

Mangiarotti, G.: Knowledge management practices and innovation propensity: a firm-level analysis for Luxembourg. Int. J. Technol. Manage. **58**(3), 261–283 (2012)

Marvel, M.R., Lumpkin, G.T.: Technology entrepreneurs' human capital and its effects on innovation radicalness. Entrepreneurship Theor. Pract. **31**(6), 807–828 (2007)

Mouritsen, J.: Problematizing intellectual capital research: ostensive versus performative IC. Acc. Auditing Accountability J. **19**(6), 820–841 (2006)

OCDE: Intellectual Assets and Innovation. The SME Dimension. OECD Studies on SMEs and Entrepreneurship. OECD Publishing (2011). http://dx.doi.org/10.1787/9789264118263-en

Pérez-Cano, C.: Firm size and appropriability of the results of innovation. J. Eng. Tech. Manage. **30**, 209–226 (2013)

Reagans, R., Zuckerman, E.W.: Networks, diversity, and productivity: the social capital of corporate R&D teams. Organ. Sci. **12**(4), 502–517 (2001)

Schumpeter, J.A.: Capitalism Socialism and Democracy. Harper and Row, New York (1950)

Steward, T.A.: Intellectual Capital. The New Wealth of Organizations. Nicholas Brealey, London (1997)

Teece, D.J.: Profiting from technological innovation: implications for integration, collaboration, licensing and public policy. Res. Policy **15**, 285–305 (1986)

Un, C.A., Cuervo-Cazurra, A., Asakawa, K.: R&D collaborations and product innovation. J. Prod. Innov. Manag. **27**(5), 673–689 (2010)

The OWA Operator with Boxplot Method in Time Series

Agustín Torres-Martínez[1(✉)], Anna M. Gil-Lafuente[1], and José M. Merigó[2,3]

[1] Department of Business, University of Barcelona, Av. Diagonal 690, 08034 Barcelona, Spain
agustoma@hotmail.com, amgil@ub.edu
[2] Department of Management, Control and Information Systems, University of Chile, Av. Paraguay 257, 8330015 Santiago, Chile
jmerigo@fen.uchile.cl
[3] School of Information, Systems and Modelling, Faculty of Engineering and Information Technology, University of Technology Sydney, 81 Broadway, Ultimo, NSW 2007, Australia
jose.merigo@uts.edu.au

Abstract. This paper presents an aggregation model for time series with the OWA operator, using as arguments representative values calculated from the quantiles of a data set and particularly with the values of the boxplot graph. The application of the OWA operator in this model allows forecasting time series including the optimistic or pessimistic attitude of the analysts, through the weighting vector. To illustrate the boxplot-OWA operator model, results are presented with data from the IBEX 35 for different periods of time in 2017.

Keywords: OWA operator · Boxplot · Weighted average · Time series

1 Introduction

Since Yager (1988) proposed the ordered weighted averaging (OWA) operator, this has been used to develop a significant number of extensions (He et al. 2017) and applications in decision-making models (Blanco-Mesa et al. 2017; Yu et al. 2016). A particular feature of the OWA aggregation models, is that it usually applies to a limited number of data or arguments that represent criteria (Casanovas et al. 2016; Dursun and Karsak 2010; Zarghami and Szidarovszky 2009; Zarghami et al. 2008) or possible scenarios for a limited number of alternatives (Casanovas et al. 2015; Merigó et al. 2014; Merigo and Gil-Lafuente 2010). This operator, when classified as an average operator, is also applicable when there is a large volume of data, in which, it may be convenient to group them before calculating measures of central tendency and dispersion. On the other hand, in the cases of analysis with a large volume of data, the definition of a weighting vector to estimate weighted averages becomes complex, since the weights can be very small.

In statistical analysis, it is usual to group the data in class intervals using frequency distributions and classifying them into different groups that are represented by class

J. C. Ferrer-Comalat et al. (Eds.): MS-18 2018, AISC 894, pp. 13–21, 2020.
https://doi.org/10.1007/978-3-030-15413-4_2

marks, which could be used as arguments to represent a data set in the aggregation of information. However, the quantiles that are measures of position (quartiles, quintiles, deciles or percentiles) are an interesting way to represent the data in a more homogeneous way and to use them as arguments in aggregation models. Therefore, from the quantiles, there are many models that can be proposed as forecasting methods for a data set, being very popular the use of quartiles and in particular the representation through the boxplot graph (Tukey 1970), that provides five representative measures of the data, which also include the minimum and maximum.

The objective of this work is to use the descriptive values, from the quantiles of a data set, and particularly with the data of the boxplot graph as representative values in aggregation models with the OWA operator for the estimation of forecasts. An interesting case is also presented with the proposed model applied to time series to forecast future values of the IBEX 35 with the daily data of 2017 for the annual, semi-annual, quarterly and monthly periods of time.

2 Preliminaries

This section briefly presents the concepts of OWA operator and boxplot graph.

2.1 The OWA Operator

The OWA operator (Yager 1988) is a type of parameterized average, in which information can be added, allowing us to underestimate or overestimate the arguments or data, through a weighting vector.

Definition 1: An OWA operator of dimension n, is a mapping $OWA : R^n \rightarrow R$ that has an associated vector W of dimension n with $w_j \in [0, 1]$ and $\sum_{j=1}^{n} w_j$. So that:

$$OWA(a_1, a_2, \ldots, a_n) = \sum_{j=1}^{n} w_j b_j \tag{1}$$

where b_j is the j-th largest of the a_j. In essence, the arguments a_j are not associated with the w_j weights directly, but with its position in the ordering.

2.2 The Ascending OWA Operator

Definition 2: An Ascending OWA (AOWA) operator is defined as a mapping of dimension n, $AOWA : R^n \rightarrow R$, that has an associated weighting vector W of dimension n, with $w_j \in [0, 1]$ and $\sum_{j=1}^{n} w_j$. So that:

$$AOWA(a_1, a_2, \ldots, a_n) = \sum_{j=1}^{n} w_j b_j \tag{2}$$

where b_j is the j-th smallest of the a_j, such that $b_1 \leq b_2 \leq \cdots \leq b_n$, which thus differs from the OWA where $b_1 \geq b_2 \geq \cdots \geq b_n$.

The difference between the OWA operator and the Ascending OWA operator is the way in which it manages the arguments, descending in the first and ascending in the second, respectively, and depends on the optimistic or pessimistic attitudes of the decision maker.

2.3 Boxplot Graphic

The boxplot graph introduced by Tukey (1970) is well known in descriptive statistics because it allows outlining the information about the position, dispersion and form of a distribution, from five statistical data that are representative: median, first quartile, third quartile, minimum and maximum. In addition, to identify atypical observations.

Definition 3: Let $X_n = \{x_1, x_2, \ldots x_n\}$ be a univariate and unimodal data set, the boxplot graph is constructed by drawing a line on the median Q_2, a box limited by the first and third quartiles Q_1 and Q_3 and two lines from the edges of the box to the values furthest from the median, without exceeding the limits of the following interval:

$$[Q_1 - 1.5IQR, Q_3 + 1.5IQR] \tag{3}$$

IQR is the interquartile range ($IQR = Q_3 - Q_1$) and the values outside the range are considered potentially as outliers.

The statistical values that are included in a boxplot graph are shown in Fig. 1.

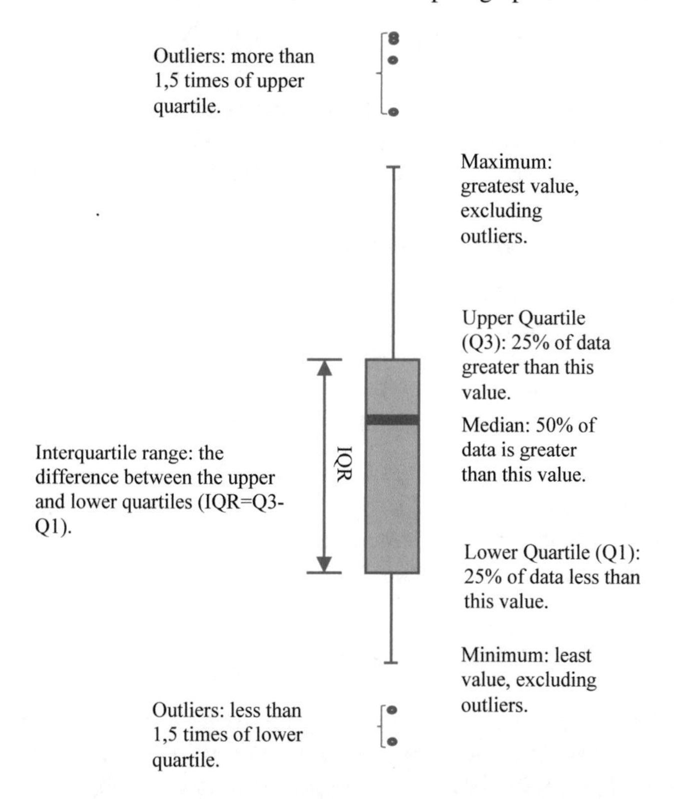

Fig. 1. Components of a boxplot graph

3 Boxplot Method with the OWA Operator

One of the main characteristics of the weighted average and particularly of the OWA operator is the use of a weighting vector, with which a different relative importance is given to each of the data. However, use a lot of data to make estimates, such as the historical information of a given event, means that the dimension of the weighting vector is equal to the volume of data to be analyzed, therefore, it is possible that many of these weights are near zero or a large amount of data being overlooked in the analysis. The model presented in this paper proposes the use of representative data in the analysis, first with boxplot graph and their respective cutoffs with or without outliers and secondly with different levels of quantiles.

Next, the proposed models are presented:

3.1 Boxplot OWA Operator

Definition 4: Let $X = \{x_1, x_2, \ldots x_p\}$ be a observations set, which are represented from a vector of dimension 5 $A = \{a_1, a_2, a_3, a_4, a_5\}$, where the values of A are the five descriptive values of the Boxplot chart and has an associated vector of dimension 5, $W = \{w_1, w_2, w_3, w_4, w_5\}$, with $w_j \in [0, 1]$ and $\sum_{j=1}^{n} w_j$. So that:

$$Boxplot - OWA(a_1, a_2, a_3, a_4, a_5) = \sum_{j=1}^{n} w_j b_j \tag{4}$$

where b_j is the largest j-th of the a_i, so the arguments are sorted in descending order. However, in this case, it is most practical to use the AOWA operator, since the data from the boxplot graph, by their nature, are presented in ascending order.

3.2 Boxplot OWA Operator with Outliers

The previous model is based on the use of five statistical data as arguments to be added. However, it is important to remember the existence of outliers, that by distancing themselves from most of the collected data they are discarded. Some of these atypical data may be useful in the analysis to assess extreme events or sudden changes in the trend of the series. Its formulation would be the following:

Definition 5: Let $X = \{x_1, x_2, \ldots x_p\}$ be a observations set, which are represented from a vector of dimension n, $A = \{a_1, a_2, \ldots, a_n\}$, where the values of A are the five descriptive values of the Boxplot chart and some outliers, which has an associated vector of dimension n, $W = \{w_1, w_2, \ldots, w_n\}$, with $w_j \in [0, 1]$ and $\sum_{j=1}^{n} w_j$. So that:

$$Boxplot - OWA(a_1, \ldots, a_n) = \sum_{j=1}^{n} w_j b_j \tag{5}$$

where b_j is the largest j-th of the a_i, so the arguments are sorted in descending order. However, in this case, it is most practical to use the AOWA operator, since the data from the boxplot graph, by their nature, are presented in ascending order.

The advantage of this second model is that using outliers, it is possible to identify extreme values.

3.3 Quantile OWA Operator

As can be seen in the boxplot OWA operator, the arguments are part of the values that classify the data in quartiles that divide the distribution into four parts. However, using the concept of statistical quantiles, it is possible to establish smaller groups, such as the case of quintiles that divide the distribution into five parts or deciles into ten parts, and so on. In this way, it can be obtained a greater number of arguments that represent the distribution for the forecast. In this case, the operator will be called from quantile established in the analysis, for example, Quintile-OWA operator or Decile-OWA operator. For a more general denomination, the quantile name will be used with the subscript corresponding to the selected level. Thus, Quintile-OWA operator is named as $Quantile_5 - OWA$ operator and Decile-OWA operator a $Quantile_{10} - OWA$.

Definition 6: Let $X = \{x_1, x_2, \ldots x_p\}$ be a observations set, which are represented from a vector of dimension n, coinciding with the quantile $n - 1$ established and $A = \{a_1, a_2, \ldots, a_n\}$, where the values of A coincide with the n descriptive values of the quantiles plus the minimum value, which has an associated vector of dimension n, $W = \{w_1, w_2, \ldots, w_n\}$ with $w_j \in [0, 1]$ and $\sum_{j=1}^{n} w_j$. So that:

$$Quantile_n - OWA(a_1, \ldots, a_n) = \sum_{j=1}^{n} w_j b_j \tag{6}$$

where b_j is the largest j-th of the a_i, so the arguments are sorted in descending order. However, in this case, it is most practical to use the AOWA operator, since the data, by their nature, are presented in ascending order.

4 Algorithm

Step 1: Choose a univariate data $C = \{c_1, c_2, \ldots c_p\}$, corresponding to the different values of a sample $X = \{x_1, x_2, \ldots x_p\}$.

Step 2: Estimate the statistical values: minimum, maximum, median, lower quartile and upper quartile and draw the boxplot chart, also identifying the outliers to analyze the convenience of including them in the model. In the case of establishing a model different from the boxplot-OWA operator, set the number of quantiles for a model quantile-OWA operator.

Step 3: Determine the arguments or representative values of the model calculated in the previous step.

Step 4: Set the weighting vector $W = \{w_1, w_2, \ldots, w_n\}$, with $n = 5$ for the Boxplot-OWA operator and $n > 5$ in cases that include outliers, and $n \neq 5$ for other models of the Quartile-OWA operator. It must be taken into account that $w_j \in [0, 1]$ and $\sum_{j=1}^{n} w_j$.

Step 5: Calculate the operator boxplot OWA operator or quantile-OWA operator, as the case may be.
Step 6: Interpret the results.

5 Application to Time Series

The proposed model can be implemented in real problems with historical information, being therefore very useful to analyze and forecast time series.

Case IBEX 35
In order to demonstrate the usefulness of the proposed model, an application with the IBEX 35 quotes is presented. For this, Fig. 2 presents the daily closing data of 2017.

Fig. 2. Daily quotes of the IBEX 35 2017

For the previous data, the following periods of time have been established: annual, II semester, III quarter and December, obtaining the boxplot shown in Fig. 3.

The statistical data for each of the selected periods are presented in Table 1.

From Table 1, the following vectors of the arguments to be added are obtained:

$$Boxplot - IBEX35_{Annual} = (9.330, 10.054, 10.305, 10.544, 11.135)$$

$$Boxplot - IBEX35_{II\ Semester} = (9.965, 10.185, 10.288, 10.450, 10.735)$$

$$Boxplot - IBEX35_{III\ Quarter} = (9.965, 10.144, 10.220, 10.305, 10.524)$$

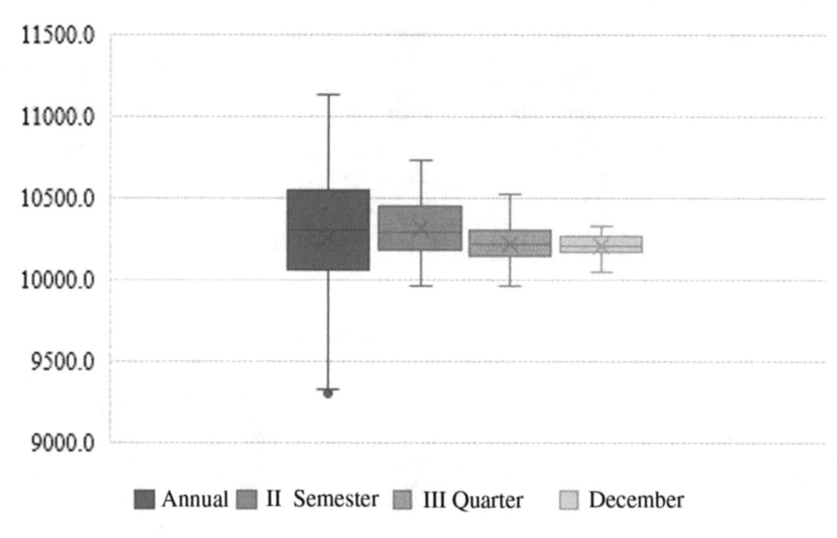

Fig. 3. Boxplot IBEX 35 2017

Table 1. Descriptions of boxplot graphics 2017

Data	Annual	II Semester	III Quarter	December
Sample	255	128	84	19
Minimum	9.330	9.965	9.965	10.044
First quartile	10.054	10.185	10.144	10.165
Median	10.305	10.288	10.220	10.208
Third quartile	10.544	10.450	10.305	10.263
Maximum	11.135	10.735	10.524	10.321
IQR	490	265	161	98
Outliers	9.305	–	–	–

$$Boxplot - IBEX35_{December} = (10.044, 10.165, 10.208, 10.263, 10.321)$$

The period selected for the analysis is of great importance because it will affect the dispersion of the values to be used as arguments. Figure 3 also shows how the data dispersion is smaller if the selected time period is smaller.

For the IBEX 35 forecast with the above data shown in Table 2 using the boxplot-OWA operator, the weighting vectors are:

The results obtained are presented in Table 3:

As can be seen in Table 3, the predictive value of this model will depend on the time period selected in the analysis and the optimistic or pessimistic attitude of the decision maker, through the weighting vector W.

Figure 4 also shows that the amplitude of the range of maximum and minimum values depends on the time period of the data. Therefore, the selection of the period will also depend on the time horizon that is intended to forecast.

Table 2. Weighting vectors

Operator	w_1	w_2	w_3	w_4	w_5
$Boxplot - OWA_{min}$	0	0	0	0	1
$Boxplot - OWA_{max}$	1	0	0	0	0
$Boxplot - OWA_{mean}$	0.2	0.2	0.2	0.2	0.2
$Boxplot - OWA_1$	0,2	0,4	0,2	0,1	0,1
$Boxplot - OWA_2$	0,1	0,1	0,2	0,35	0,25

Table 3. Aggregated results

Time frame	$Boxplot - OWA_{min}$	$Boxplot - OWA_{max}$	$Boxplot - OWA_{mean}$	$Boxplot - OWA_1$	$Boxplot - OWA_2$
Annual	9.330	11.135	10.274	10.117	10.474
II Semester	9.965	10.735	10.325	10.243	10.414
III Quarter	9.965	10.524	10.232	10.178	10.293
December	10.044	10.321	10.200	10.175	10.235

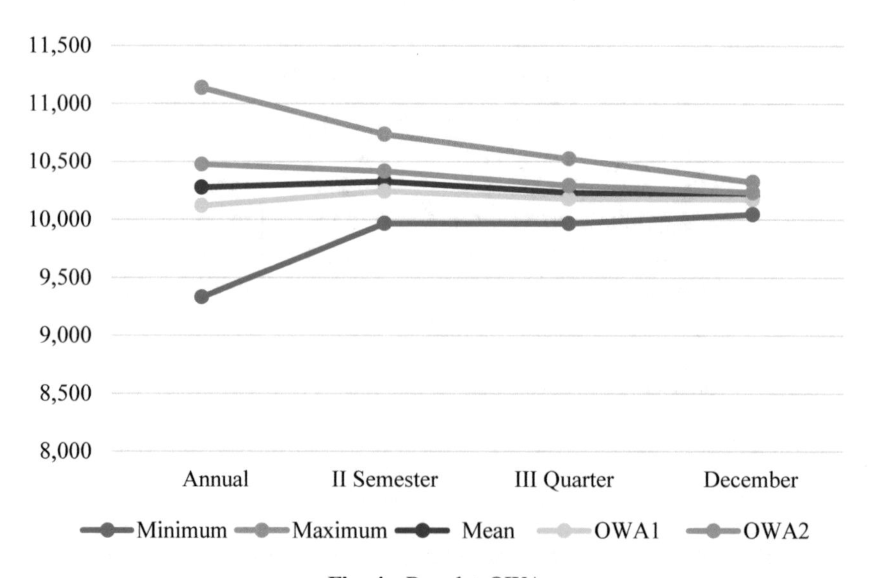

Fig. 4. Boxplot OWA

6 Conclusions

In this paper, we present a new model of OWA operators using quantiles as arguments, in cases where there is a lot of data and it would be complex to assign a significant weight to each value. The advantage of this method is that from a large volume of data it is possible to limit the arguments to be added to some key points of a given distribution, being the quantiles and particularly the quartiles with the use of the boxplot chart an interesting method to represent the information. The use of the OWA operator in this model, either using as representative data the information of the boxplot chart or from the calculation of n quantiles and/or outliers, it is an opportunity to include the optimistic or pessimistic attitude of the decision makers in the forecasts. However, the results obtained with the proposed model do not pretend to be an alternative to the statistical methods of time series forecasts, although it does offer a different forecast that can be considered by decision makers in situations of great uncertainty.

References

Blanco-Mesa, F., Merigó, J.M., Gil-Lafuente, A.M.: Fuzzy decision making: a bibliometric-based review. J. Intell. Fuzzy Syst. **32**(3), 2033–2050 (2017)

Casanovas, M., Torres-Martínez, A., Merigó, J.M.: Decision making processes of non-life insurance pricing using fuzzy logic and OWA operators. Econ. Comput. Econ. Cybern. Stud. Res. **49**(2), 169–187 (2015)

Casanovas, M., Torres-Martínez, A., Merigó, J.M.: Decision making in reinsurance with induced OWA operators and Minkowski distances. Cybern. Syst. **47**(6), 460–477 (2016)

Dursun, M., Karsak, E.E.: A fuzzy MCDM approach for personnel selection. Expert Syst. Appl. **37**(6), 4324–4330 (2010)

He, X., Wu, Y., Yu, D., Merigó, J.M.: Exploring the ordered weighted averaging operator knowledge domain: a bibliometric analysis. Int. J. Intell. Syst. **32**(11), 1151–1166 (2017)

Merigó, J.M., Casanovas, M., Xu, Y.: Fuzzy group decision-making with generalized probabilistic OWA operators. J. Intell. Fuzzy Syst. **27**(2), 783–792 (2014)

Merigo, J.M., Gil-Lafuente, A.M.: New decision-making techniques and their application in the selection of financial products. Inf. Sci. **180**(11), 2085–2094 (2010)

Tukey, J.W.: Exploratory Data Analysis. Addison-Wesley, Reading (1970)

Yager, R.: On ordered weighted averaging aggregation operators in multicriteria decision-making. IEEE Trans. Syst. Man Cybern. **18**(1), 183–190 (1988)

Yu, D., Li, D.-F., Merigó, J.M., Fang, L.: Mapping development of linguistic decision making studies. J. Intell. Fuzzy Syst. **30**(5), 2727–2736 (2016)

Zarghami, M., Szidarovszky, F.: Revising the OWA operator for multi criteria decision making problems under uncertainty. Eur. J. Oper. Res. **198**(1), 259–265 (2009)

Zarghami, M., Szidarovszky, F., Ardakanian, R.: A fuzzy-stochastic OWA model for robust multi-criteria decision making. Fuzzy Optim. Decis. Making **7**(1), 1–15 (2008)

Weighted Logistic Regression to Improve Predictive Performance in Insurance

Jessica Pesantez-Narvaez(✉) and Montserrat Guillen

Department of Econometrics, Riskcenter-IREA, University of Barcelona,
Av. Diagonal 690, 08034 Barcelona, Spain
{jessica.pesantez,mguillen}@ub.edu

Abstract. We propose a logistic regression model combined with a weighting estimation procedure that incorporates a tuning parameter. We analyse predictive performance indicators. Results show that the parameter defining the weights can be used to improve predictive accuracy, at least when the original predictive value is distant from the response average. We use a publicly available data set to illustrate our method and we discuss the potential benefits of this methodology in the decision to purchase full coverage motor insurance versus a basic insurance product.

Keywords: Maximum weighted likelihood · Extremes · Confusion matrix · Binary choice

1 Introduction

Predictive models in the insurance industry are used to support management decisions mainly related to risk measurement and prevention. The scope of the predictive analytics is even broader in areas like sales and marketing to anticipate customers' backlashes, or in compensation analysis to reward particular employees' behavior, or in finance to forecast financial results (Frees et al. 2014). Accurate predictions are needed, to assess risk, to calculate the price of insurance policies, to forecast future liabilities or to detect insurance fraud.

One popular method for binary dependent variables is the logistic regression model. This is implemented in a sample of n individuals who might correspond to policyholders, firms, business units or agents. The variable of interest is a binary response and every observation has a set of K covariates or characteristics that influence the dependent variable. The logistic regression final result allows predicting the expected value of the response, so the probability of occurrence.

Initially, all observations have the same weight. This means that the observations have exactly the same relevance for the inference, or it is assumed that the observations are generated by a simple random sample. So, each observation has the same degree of representativeness of the population from which it has been extracted. The idea to introduce weights in the likelihood estimation is considered to improve predictive accuracy. In fact, this approach was firstly introduced in the missing data literature by Robins et al. (1994). In geography studies, where the effect of each observation in the

J. C. Ferrer-Comalat et al. (Eds.): MS-18 2018, AISC 894, pp. 22–34, 2020.
https://doi.org/10.1007/978-3-030-15413-4_3

estimation result might strongly depend on the location (Agterberg et al. 1993), weights are used to correct this effect.

Sampling weighting was initially motivated to approximate the distribution of the population. In statistics, in the univariate setting a weighted average is calculated by multiplying the weighting W_i by the value of each observation (Winship and Radbill 1994) and use the sum of weights is equal to sample size. Note that weights are strictly positive numbers. They are interpreted as a measure of the size, or relative size, of each observation with respect to the number of population units that this observation represents.

Weighting can be aimed at differentiating the contribution of observations, so an individual, denoted by subscript *i*, with a small weight, it will have little influence on the results compared to other observations with a larger weight.

Our objective is to study the influence of changing the weights of the initial sample in order to improve the accuracy of a simple predictive model. In particular, we focus on the logistic regression model and we study extreme observations with respect to the covariates, the ones that are farthest from the corresponding mean. As these data points might be considered unusual, their probability might be low and the predictive model can be inaccurate to anticipate their response. Our aim is to improve the prediction for this particular part of the sample.

We propose a special weighting procedure to be incorporated in the likelihood estimation of a logistic regression as a particular case of the weighted likelihood method. The definition of the weights depends on a parameter which we call the *tuning parameter*. A discussion on the sensibility of precisely this tuning parameter is presented. Moreover, we address how to find an optimal value for the tuning parameter. We use one public data set that contains a real sample of insurance customers' information and their decision to buy a full coverage insurance versus a basic insurance product (Guillen 2014). All analyses are performed in R language.

This paper is divided in the following four parts. In Sect. 2, we present the methodology description where the role of the proposed tuning parameter is discussed. Section 3 describes the data and the results of the performance measures which are used to evaluate the sensitivity reaction in predictive capacity of the weighted logistic regression model. Section 4 contains final conclusions.

2 Methodology Description

Let us assume that our response variable, $\boldsymbol{Y}_i$, $i = 1,\ldots, n$ has been observed and that $\mathbf{X}_{ik}$, $k = 1, \ldots, K$ is the corresponding set of covariates. In this article we assume that $\boldsymbol{Y}_i$ takes two possible values, which we consider coded as 1 for the occurrence of the event and 0 for the non-event.

2.1 Logistic Regression

A logistic regression model is a commonly used regression method to predict a binary discrete choice endogenous variable explained by one or more nominal, ordinal or ratio-level exogenous variables (Greene 2002). Additionally, it is a particular predictive modelling technique because it aims at finding the probability of occurrence of an event and the result is bounded between 0 and 1.

The logistic regression model is a particular case of the generalized linear model (McCullagh and Nelder 1989). The logit function is the canonical link and is given as:

$$g(\Pi_i) = log\left(\frac{\Pi_i}{1-\Pi_i}\right) = \beta_0 + \sum_{k=1}^{K} X_{ik}\beta_k, \tag{1}$$

for $i = 1,\ldots, n$ observations, $k = 1,\ldots, K$ denoting the covariates where $\beta_0, \beta_1, \ldots, \beta_k$ are the model parameters, and Π_i is the probability of the observed event in response Y_i. Thus, in the logistic regression model, the logit transform, i.e. the log-odds of the probability of the event, equals the linear predictor:

$$P(Y_i = 1) = E(Y_i) = \frac{e^{(\beta_0 + \sum_{k=1}^{K} X_{ik}\beta_k)}}{1 + e^{(\beta_0 + \sum_{k=1}^{K} X_{ik}\beta_k)}}. \tag{2}$$

A logistic regression can be estimated by maximum likelihood (for further details see McCullagh and Nelder 1989).

2.2 The Weighted Likelihood Estimation and the Tuning Parameter as a Weighting Mechanism

To formally define the concept of weighted logistic regression, we first address the notion of weighted likelihood estimation in general.

Let $\tilde{Y} = (Y_1, Y_2, \ldots Y_n)'$ be a simple random sample of a binary random variable with a probability density function:[1]

$$\mathrm{P}(Y_i = y_i|X_i, \beta) = \frac{e^{y_i(\beta_0 + \sum_{k=1}^{K} X_{ik}\beta_k)}}{1 + e^{(\beta_0 + \sum_{k=1}^{K} X_{ik}\beta_k)}}, \tag{3}$$

with the vector of parameters $\beta = (\beta_0, \beta_1, \beta_2, \ldots, \beta_K{}')$ where $\beta \subseteq \mathbb{R}^{K+1}$.

Let χ be the sampling space, in other words, all possible values of $\tilde{Y}$. Then the likelihood function is defined for $\tilde{Y} = (Y_1, Y_2, \ldots Y_n)' \in \chi$ as:

$$\begin{aligned} \mathbf{L}\,(*|\tilde{Y}) : \beta \rightarrow L(\beta|\tilde{Y}) &= P(\tilde{Y}|X, \beta) \\ &= \prod_{i=1}^{n} \mathrm{P}(Y_i = y_i|X_i, \beta) \end{aligned} \tag{4}$$

[1] We use ’ to denote the transpose of a vector.

We take the logarithm of the likelihood because it is a strictly increasing function and the extreme points are the same for the logarithm of the likelihood function and for the likelihood function itself. Consequently, using the properties of the logarithm, we define the log-likelihood function as $\ln L(\beta|\tilde{Y}) = \sum_{i=1}^{n} ln\, P(Y_i = y_i|X_i, \beta)$.

For each $\tilde{Y} \in \chi$, the maximum likelihood estimator of β is denoted as $\hat{\beta}$ and it corresponds to the value of β that maximizes the likelihood $\mathbf{L}\,(*|\tilde{Y})$ since the purpose is to find the parameters for which the probability of the observed data is the greatest possible value:

$$\mathbf{L}(\hat{\beta}|\tilde{Y}) = max_{\beta} \ln \prod_{i=1}^{n} f(Y_i|\beta). \tag{5}$$

The weighted logistic regression method is based on a weighted log-likelihood estimation, denoted by $\boldsymbol{l}(\beta)$, which is defined as:

$$\boldsymbol{l}(\beta) = \sum_{i=1}^{n} W_i * ln\; P(Y_i = y_i|X_i, \beta), \tag{6}$$

where $W_i = (W_1, W_2, \ldots, W_n)'$ is the vector of weights.

This modification can be a consequence, for instance, of the existence of common *sampling weighting* $\underline{Wi} = Q_i/H_i$ where Q_i is the fraction of the decision-making population, and H_i is the analogous fraction of the decision-making sample that are represented by observation i (Manski and Lerman 1977).

In this paper, we propose a vector of weights which is constructed as follows:

$$W_i = |\hat{y}_i - \bar{y}|^{\theta}, \quad i = 1\ldots, n \tag{7}$$

where $\hat{y}_i$ is defined as the estimated probability obtained by the standard logistic regression model for observation i (as in (2) from Sect. 2.1). Let us consider $\bar{y}$ as the mean value of the endogenous variable. The weights' definition depends on θ, which is a real number and can be called the tuning parameter. This parameter is calibrated later[2], however it should be noted that when $\theta = 0$, then all weights are equal to one.

A change of the vector of weights determines a change in the estimated model coefficients, as well as their level of significance, which can even reverse the impact from positive to negative or the other way round, the weighted estimation procedure can modify the magnitude of influence on the outcome binary variable of each covariate, which, in turn, directly influences the results and so, the confusion matrix.

The main idea for defining weights as (7) is to find the best tuning value. These weights depend on the distance between the initial predictive value and the mean of the observed outcome. For a positive tuning parameter the weight is larger in the most

[2] The tuning parameter is chosen so that the best value provides the best predictive performance obtained by the goodness of fit tests.

extremal observations which lay far from the mean, whereas it is smaller more than those that are close to the mean. The possible scenarios for selecting the tuning parameter are:

$\theta = 0$;	The maximum likelihood estimation remains the same as the unweighted model,
$\theta > 0$;	The weighting gives more importance to the observations whose original predictive value is far from the mean,
$\theta < 0$;	The weighting gives more importance to the observations whose original predictive value is close to the mean

In this paper, the concept of weight is not defined as in other cases in the literature on weighted regression. Other approaches such as Adaboost and similar machine learning algorithms (see Friedman et al. 2000) have a totally different approach to weighting. In that case, more weight is given to wring predictions and less weight is given to correct predictions.

In our proposal, we do not look at the similarity between the predicted and the observed response. Observations that are distant from the average predicted response should be given more importance than to those that are closer to the average.

Our proposal is linked to the different concepts around the notion of an outlier observation. Indeed, we define weights mainly (although not exclusively) depending on covariates, through predicted responses.

We do not correct the estimation procedure directly in order to improve accuracy in one step. Our idea is to look at the distance between the predicted value and the observed value for each observation and then to re-estimate. This difference is substantial and it is the reason why our contribution differs, up to our knowledge, to existing approaches.

The notion of Real Adaboost, coined by Friedman et al. (2000), suggests that the weight should be a transformation of the class probability estimate. These authors show the statistical equivalence of the weighted estimating procedures to the minimization of a loss functions. Our proposal is an additional approach that is suitable for distant observations, where distance is defined by a norm in the space of the covariates.

3 Data

Data have been taken from a Spanish insurance company. A sample of 4,000 policy holders of motor insurance has been analyzed.[3] The database contains seven variables which are described in Table 1.

Additionally, Table 2 shows some brief descriptive statistics of the data. Firstly, the percentage of women who purchase a full coverage insurance vs a basic coverage is quite similar (almost a half) whereas 70.27% men decided to buy a basic coverage product. Furthermore, a big percentage of the married insurance holders seem to prefer the basic coverage with reference to the single and other insurance holders. Most

[3] The data set can be found in the following web of R resources for quantitative analysis at the University of Barcelona: www.ub.edu/rfa/R.

Table 1. Description of the variables in the insurance purchase decision data set

Type	Variable	Description
Dependent	Full coverage	Binary variable which takes the value of 1 if the policy holder decision is to purchase full coverage insurance. And 0, if he/she wants to purchase a basic coverage insurance
Independent	Age	Continuous numeric variable that represents the policy holder's age
	Seniority	Continuous numeric variable that represents the seniority (years) in the company
	Men	Binary variable that takes the value of 1 if the policy holder is a man or 0 if woman
	Urban	Binary variable that takes the value of 1 if the policy holder usually drives in an urban area, and 0 in a rural area
	Private	Binary variable that takes the value of 1 if the vehicle has a private use, and 0 if it has a commercial use
	Marital	Categorical variable that takes three states of marital status: single, married and others

Source: Guillen (2014) available at ww.ub.edu/rfa/R/regression_with_categorical_dependent_variables.html

Table 2. Insurance purchase decision descriptive statistics

Variables		Basic coverage (y = 0)	Full coverage (y = 1)	Total
Age (years)		48.27	43.09	46.47
Seniority in company (years)		9.93	12.66	10.88
Sex	Woman	498 (50.30%)	492 (49.70%)	990
	Man	2115 (70.27%)	895 (29.73%)	3010
Driving area	Rural	1906 (72.83%)	711 (27.17%)	2617
	Urban	707 (51.12%)	676 (48.88%)	1383
Vehicle use	Commercial	33 (84.62%)	6 (15.38%)	39
	Private	2580 (65.14%)	1381 (34.86%)	3691
Marital Status	Single	467 (54.24%)	394 (45.76%)	861
	Married	2047 (68.85%)	926 (31.15%)	2973
	Other	99 (59.64%)	67 (40.36%)	166

Note: Continuous variables are expressed in the mean. The number of observations is 4,000. The percentage of policy holders who choose full coverage is 34.68%.

people who drive in rural areas have purchased a basic coverage while people in urban areas have almost a similar tendency between basic and full coverage. Moreover, the average age of insurance holders who choose a basic coverage is older than the one of full coverage. And finally, people who have more seniority in the company purchase more often full coverage insurance than newer customers.

4 Results

We want to evaluate the decision to purchase a full coverage insurance (coded as 1) versus a basic coverage (coded as 0) determined by some exogenous variables through a logistic regression model:

$$\mathrm{Prob}(Y_i = 1) = \left(\frac{e^{z_i}}{1 + e^{z_i}}\right) \tag{8}$$

$$\mathrm{Prob}(Y_i = 0) = 1 - \left(\frac{e^{z_i}}{1 + e^{z_i}}\right) \tag{9}$$

$$\begin{aligned} z_i = &\beta o + \beta 1 * Age_i + \beta 2 * Seniority_i + \beta 3 * Men_i \\ &+ \beta 4 * Urban_i + \beta 5 * Private_i + \beta 6 * Marital_i, \end{aligned} \tag{10}$$

where Y_i is the response variable, in this case the binary full coverage purchase variable, βo is the constant coefficient, and β_k are the coefficients related to the independent variables in Table 2. Similar approaches with discussions on the classifiers can be found in Guelman and Guillen (2014), the decision to purchase a certain type of insurance together with the price setting is discussed in Guelman et al. (2014; 2015a, b).

In this section, we study the tuning parameter behavior in the proposed weighting procedure through some statistical measures. In the appendix we present the model estimates for the classical logistic regression (unweighted) and the weighted logistic regression with a tuning parameter equal to 1.

4.1 Weighted Log-Likelihood Performance

The log-likelihood function summarizes information on the parameter that is given by the sample. Since the original likelihood estimation is now being modified by the proposed weights, it is necessary to ensure that the new function is still concave. So, a global maximum likelihood estimate can be found numerically after a few iterations.

Figure 1 shows the maximum log likelihood values with $\theta \in \{0, 30\}$ where a maximum of all can be detected when $\theta \in \{7, 10\}$.

4.2 Norm of the Estimated Parameters

A norm, denoted by $\|\cdot\|$, finds a strictly positive length of a vector V in a vector space $(V, \|\cdot\|)$. The norm metric on V is generally defined by $\|z - q\|$ with z and q vectors (Deza and Deza 2009).

The intuitive idea of measuring the distance between the estimated coefficients $\hat{\beta}$ from the base model (2) and the estimated coefficients $\hat{\beta}_\theta$ from the proposed weighted logistic model with the weighting procedure of (7) is defined as the Euclidean distance between vectors, namely, $\left\|\hat{\beta} - \hat{\beta}_\theta\right\|^{1/2}$ with $\theta \in \{0, 10\}$.

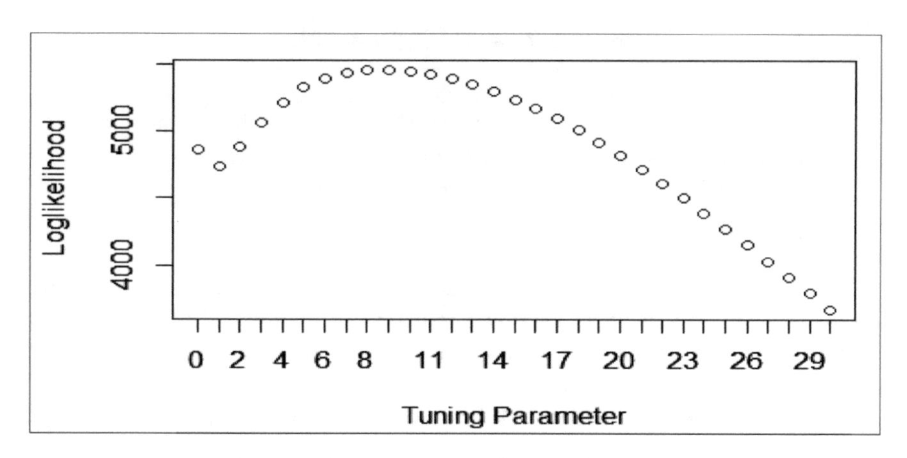

Fig. 1. Maximum log likelihood values of the estimated models versus the tuning parameter from 0 to 30.

Figure 2 shows that the norm when $\theta \in \{2, 4\}$ is approximately the largest which means that the tuning parameter taken in this interval already shows that the parameter estimates for the weighted maximum likelihood are distant from the unweighted model.

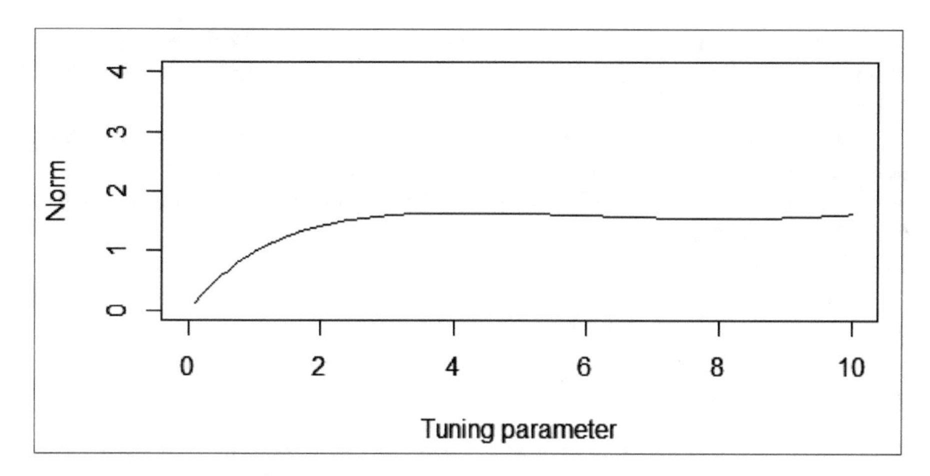

Fig. 2. Norm of the difference between the vector of estimated parameters in the unweighted model and the vector of estimated parameters in the weighted model

4.3 Confusion Matrix

A confusion matrix or classification matrix is mainly used to measure the classification performance of classifiers with respect to some test data in studies of artificial intelligence, information retrieval and data mining (Jiang and Liu 2013). Thus, this predictive method is used to evaluate the accuracy of the results of the model under a given classifier (Ting 2017).

Table 3. Confusion matrix definition

		Predicted	
		Basic coverage (y = 0)	Full coverage (y = 1)
Observed	Basic coverage (y = 0)	True negative *(TN)*	False positive *(FP)*
	Full coverage (y = 1)	False negative *(FN)*	True positive *(TP)*

A confusion matrix is a two-dimensional matrix where observed data is compared with the predicted values under the given classification algorithm.[4] Table 3 shows the four alternative classification outcomes when placing the models results into a confusion matrix.

The three measures of classification performance that we are going to analyze are:

- Sensitivity, which measures the proportion of policy holders that were classified in the full coverage insurance among those who effectively purchased full coverage insurance. TP/(TP + FN)
- Specificity, which measures the rate between the policy holders who were classified in the basic coverage insurance among those who purchased basic coverage insurance. TN/(TN + FP)
- Accuracy, which measures the rate of policy holders who are correctly classified. $(TP + TN)/(TP + TN + FP + FN)$.

Consequently, the confusion matrix is used to measure the predicting performance of a model with θ varying from to 0 to 10. The purpose is to find the value of θ that guarantees the highest levels of sensitivity, specificity and accuracy.

In Fig. 3 the tuning parameter value of each model is written above each plotted point. Sensitivity and specificity are evaluated at a threshold equal to 0.3.

The purpose of Fig. 3 is to find the model that is geometrically closest[5] to the point (0,1). This rule is considered as the optimal criterion to find the best predictive model with high sensitivity and high specificity.

The optimal tuning parameter θ is equal to 1 and, in this case the sensitivity is 0.84, the specificity is 0.66 and the accuracy is 0.72.

4.4 Extreme Points Analysis

Data points which are far from the mean predictors can be considered extreme. Thus, the first and the last decile of the predictions (associated to policy holders who are the least likely to purchase a full coverage insurance and the policy holders who are the most likely to purchase a full coverage insurance respectively) are analyzed.

The root mean square error (RMSE) is calculated for the first and the last decile of predictions from the weighted ($\theta = 1$) and the unweighted logistic regression models

[4] Classifiers are rules the supervised learning like regression with an algorithm to solve classification maximizing the accuracy or reducing the error rate and keeping the training and test data the same (Lanzi 2017).

[5] It was measured with the Euclidean distance.

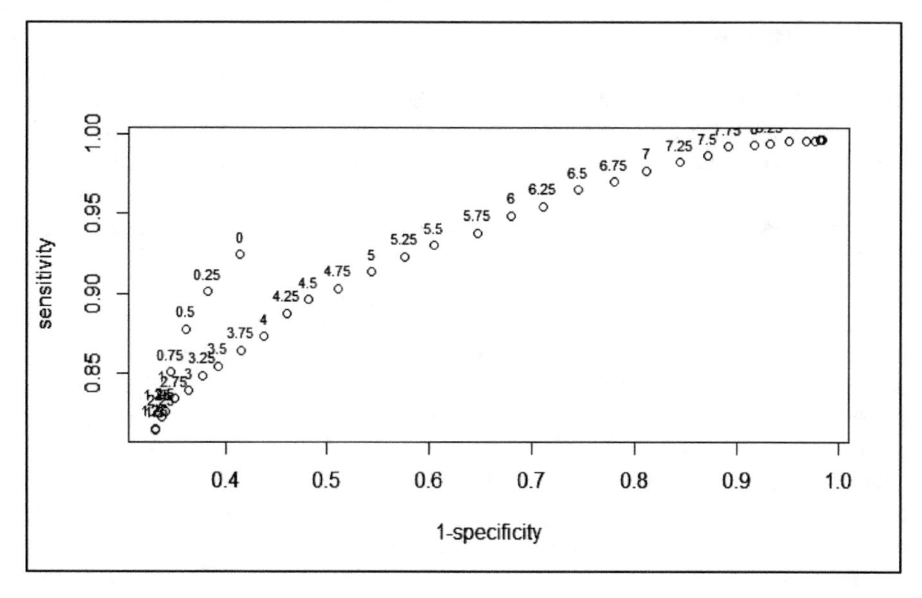

Fig. 3. Classification performance (sensibility and 1-specificity) of the estimated weighted logistic regressions.

($\theta = 0$). The RMSE is used to measure the distance between the predicted values by the model are from the observed ones. Then the smaller RMSE value the model has, the better predictive performance it has (Surhone et al. 2010).

The RMSE is defined after a table has been constructed to separate the observations in deciles according to the predictive value. Then,

$$\text{RMSE} = \sqrt{\sum\nolimits_{p=1}^{N} \frac{\left(y_p - \hat{y}_p\right)^2}{N}}, \tag{11}$$

where N is the number of observations in each decile, in this case 400.

Table 4 shows that the weighted logistic regression model has a lower RMSE in the highest decile of predictions. This model has a predictive accuracy that is better than an unweighted model for those policy holders with a large predicted probability of purchasing full coverage insurance. The RMSE is also small in the smallest decile.

Table 4. The root mean squared error for deciles of predictions in the insurance purchase data set

	Smallest decile	Highest decile
Unweighted logistic regression model	0.492	3.161
Weighted logistic regression model ($\theta = 1$)	0.385	1.511

5 Conclusions

Based on the first exploratory analysis from Sect. 4.1, the maximum log-likelihood values among all the estimated weighted logistic regression models correspond to a tuning parameter between 7 and 10. The intuitive idea is that, in the weighting mechanism, a tuning parameter can improve the likelihood of the estimated model.

The results show that a tuning parameter between 2 and 4 has the largest norm of the difference between the vector of estimated parameters in the weighted model and the vector of estimated parameters in the unweighted model.

The best tuning parameter that accomplishes the highest specificity and sensitivity rates is equal to 1. This choice is based on the optimal criterion presented in Sect. 4.3. For this case, the estimated weighted model has less root mean squared error in the extreme deciles than the unweighted model. The proposed weighting mechanism obtains more correct predictions than the classical logistic regression for the policy holders that are most likely to purchase full coverage insurance. Thus, future retention managerial strategies for this group of insurance policy holders can be based on the proposed weighted estimation procedure with $\theta = 1$. Our conclusion is that weighted logistic regression offers an array of opportunities to improve classifiers and we aim at pursuing further research in the analysis of subsamples of the population that correspond to the extremes, rather than looking at the global performance.

Acknowledgements. We thank the Spanish Ministry of Economy, FEDER grant ECO2016-76203-C2-2-P.

Appendix

Table A1. Parameter estimates for the unweighted logistic regression model in the insurance purchase data set

Variable	Parameter estimate	P-value
(Intercept)	–0.257 (0.486)	0.5959
Men	–0.961 (0.086)	0.001
Urban	1.173 (0.078)	0.001
Private	1.065 (0.469)	0.0232
Marital (married)	–0.083 (0.096)	0.3839
Marital (others)	0.161 (0.200)	0.4212
Age	–0.058 (0.004)	0.001
Seniority	0.133 (0.007)	0.001

The standard errors are expressed in brackets

Table A2. Parameter estimates for the weighted logistic regression model (θ = 1)

Variable	Parameter estimate	P-value
(Intercept)	–0.479 (0.527)	0.3632
Men	–0.782 (0.109)	0.001
Urban	0.908 (0.099)	0.001
Private	1.059 (0.495)	0.033
Marital (married)	–0.020 (0.124)	0.869
Marital (others)	–0.236 (0.253)	0.351
Age	–0.048 (0.005)	0.001
Seniority	0.099 (0.008)	0.001

The standard errors are expressed in brackets

Table A3. Confusion matrix for the unweighted logistic regression model in the insurance purchase data set

		Predicted	
		Basic coverage (y = 0)	Full coverage (y = 1)
Observed	Basic coverage (y = 0)	1708	905
	Full coverage (y = 1)	171	1216

Table A4. Confusion matrix for the weighted logistic regression model in the insurance purchase data set with a tuning parameter equal to 1.

		Predicted	
		Basic coverage (y = 0)	Full coverage (y = 1)
Observed	Basic coverage (y = 0)	1729	226
	Full coverage (y = 1)	884	1161

References

Agterberg, F.P., Bonham-Carter, G.F., Cheng, Q., Wright, D.F.: Weights of evidence modeling and weighted logistic regression for mineral potential mapping. In: Davis, J.C., Herzfeld, U.C. (eds.) Computers in Biology-25 Years of Progress, pp. 13–32. Oxford University Press, New York (1993)

Deza, M.M., Deza, E.: Encyclopedia of distances. In: Encyclopedia of Distances. Springer, Heidelberg (2009)

Frees, E.W., Derrig, R.A., Meyers, G.: Predictive Modeling Applications in Actuarial Science, vol. 1. Cambrigde University Press, Cambrigde (2014)

Friedman, J., Hastie, T., Tibshurani, R.: Additive logistic regression: a statistical view of boosting. Ann. Stat. **28**(2), 337–407 (2000)

Greene, W.: Econometric Analysis. Prentice Hall, New York (2002)

Guelman, L., Guillen, M.: A causal inference approach to measure price elasticity in automobile insurance. Expert Syst. Appl. **41**(2), 387–396 (2014)

Guelman, L., Guillen, M., Pérez-Marin, A.M.: A survey of personalized treatment models for pricing strategies in insurance. Insur. Math. Econ. **58**(1), 68–76 (2014)

Guelman, L., Guillen, M., Pérez-Marin, A.M.: Uplift random forests. Cybern. Syst. **46**(3–4), 230–248 (2015a)

Guelman, L., Guillen, M., Pérez-Marin, A.M.: A decision support framework to implement optimal personalized marketing interventions. Decis. Support Syst. **72**, 24–32 (2015b)

Guillen, M.: Regression with categorical dependent variables. In: Frees, E.W., Derrig, R.A., Meyers, G. (eds.) Predictive Modeling Applications in Actuarial Science, pp. 65–86 (2014)

Jiang, N., Liu, H.: Understand system's relative effectiveness using adapted confusion matrix. In: Marcus, A. (ed.) Design, User Experience and Usability Design Philosophy, Methods, and Tools. DUXU203. Lecture Notes in Computer Science, Springer, Heidelberg (2013)

Lanzi, P.L.: Classifier Systems. In: Sammut, C., Webb, G.I. (eds.) Encyclopedia of Machine Learning and Data Mining. Springer, Boston (2017)

Manski, C.F., Lerman, S.R.: The estimation of choice probabilities from choice based samples. Econometrica, 1977–1988 (1977)

McCullagh, P., Nelder, J.: Generalized Linear Model. Chapman and Hall/CRC, Boca Raton (1989)

Winship, C., Radbill, L.: Sampling Weights and Regression Analysis. Sociol. Methods Res. **23**(2), 230–257 (1994)

Robins, J.M., Rotnitzky, A., Ping, L.: Estimation of regression coefficients when some regressors are not always observed. J. Am. Stat. Assoc. 846–866 (1994)

Surhone, L.M., Timpledon, M.T., Marseken, S.F.: Root Mean Square Deviation. Betascript Publishing (2010)

Ting, K.M.: Confusion Matrix. In: Sammut, C., Webb, G.I. (eds.) Encyclopedia of Machine Learning and Data Mining. Springer, Boston (2017)

Light Innovation in Energy Supply for Non-connected Areas in Colombia: Partial Research via the Forgotten Effects Model

Carlos Ariel Ramírez-Triana(✉), Maria Alejandra Pineda-Escobar, Mauricio Alejandro Cano-Niño, and Sergio Alberto Mora-Pardo

Institución Universitaria Politécnico Grancolombiano, Facultad de Negocios, gestión y sostenibilidad, Calle 57 no 3 - 00 este, Bogotá, Colombia
carlos.ramirez.triana@gmail.com

Abstract. This research study proposes the analysis of the causes for the lack of technological development in Non-Interconnected Zones (ZNI) to the National Electrical Grid in Colombia, and their corresponding consequences. The bridge that connects the elucidation of this problem in terms of its causes and consequences is materialized through the application of the Theory of Forgotten Effects based on fuzzy logic.

Keywords: Theory of Forgotten Effects · Fuzzy logic · Causes

1 Introduction

The energy requirements of a country along with their supply method, determine the country's possibilities of development to a large extent. However, it is not enough to simplify the problem to a balance between energy supply and demand. Instead, an analysis that considers the country's type of resources and identifies strengths on several fronts, would be required. As well as the detection of alerts that can contribute to address the gaps faced by some sectors of society through the refinement of policies issued by decision makers.

Two acute weaknesses have been identified in the Colombian energy sector. On the one hand, climate change events such as El Niño put the country at risk of not being able to respond efficiently to a safety situation in the system. On the other hand, Colombia has a problem of electricity supply coverage in remote communities that are not connected to the national network. This research focuses on the latter weakness.

The National Interconnected System (SIN) in Colombia covers approximately 96.6% of the population in 48% of the country's territory. In remote communities, there are around 500000 households with precarious or non-existent access to electricity. Locations that are not connected to the national electricity grid are known as Non-Interconnected Zones (ZNI). The provision of electricity in those cases would rely on other sources such as diesel power plants with high environmental, financial, and social costs. In addition, there are significant geographical and technical barriers to the provision of these services in remote areas. Increasing access to electricity in Colombia's

J. C. Ferrer-Comalat et al. (Eds.): MS-18 2018, AISC 894, pp. 35–50, 2020.
https://doi.org/10.1007/978-3-030-15413-4_4

ZNIs is a challenge that must be addressed by the corresponding authorities and the affected communities, by providing sufficient, efficient, and continuous supply of energy.

Based on the Theory of Forgotten Effects, this research study attempts to analyze the causes for the absence of technological development in Colombian ZNIs, as well as their corresponding consequences. The Theory of Forgotten Effects (Kaufmann and Gil-Aluja 1988) corresponds to an application of fuzzy logic under the so-called fuzzy subsets. Based on expert analysis, the model is based on the weighting of cause–effect relationships among the variables that each expert has assessed, which allows for the identification of obvious effects that have been affecting a particular problem.

This research will use the FuzzyLog software[1] to establish the forgotten effects to address the problem of lack of electricity supply in non-connected areas in Colombia.

The paper is organized as follows. A general framework of the national energy structure will be provided in the second section, together with a more detailed reflection on the coverage level in the electrical grid, to identify current ZNIs and the ways in which, from an institutional point of view, this problem may be addressed. After that, the Forgotten Effects Model is explained in the third section, and the research advances are shown in the fourth section, determining the causality relationships that have presumably been ignored in the analysis of the studied problem. The paper finalizes by presenting the discussion of the results and the conclusion.

2 General Context of the Non-interconnected Zones in Colombia

Colombia is a country located in the Northwestern region of South America. Its population is close to 49 million inhabitants. Its territorial extension is 1,141,748 km^2. The country has access to the Atlantic and Pacific Oceans with a maritime extension of approximately 988,000 km^2.

According to the World Bank, Colombia is a medium–high income country, considered as one of the economies with greater strength and dynamism in Latin America (OECD 2013), and whose average annual growth during the last decade was 4.7%. According to the 2016 Human Development Report, Colombia is classified among the countries with high human development, with a value of 0.727 and occupying the 95th position among the 188 countries included in the index (UNDP, 2016). This figure is below the average for the Latin American region (0.748) and much lower than the average reached by OECD countries (0.880). Out of a total population of 47 million inhabitants, more than 32% live in poverty and 10% live in extreme poverty (Pineda-Escobar 2015). This proportion shows that inequality in the distribution of income continues to be an acute problem in the country, leading Colombia to be one of the countries with one of the highest Gini coefficients, not only in Latin America but also worldwide (OECD 2013). As of 2015, with an index of 53.5, Colombia is the second most inequitable country in the American continent, only behind Honduras.

[1] http://fuzzyeconomics.com/fuzzylog/index.php.

Regarding energy power, however, Colombia is a self-sufficient country. According to the information published by the International Energy Agency for 2014, the Total Primary Energy Supply[2] reached 34,008 ktoe (thousands of tons of oil equivalent) or 1.42 EJ (See Fig. 1). This result comes from a growing trend that has been developing for more than 4 decades (with a small interruption by the end of the 90 s) (IEA 2016).

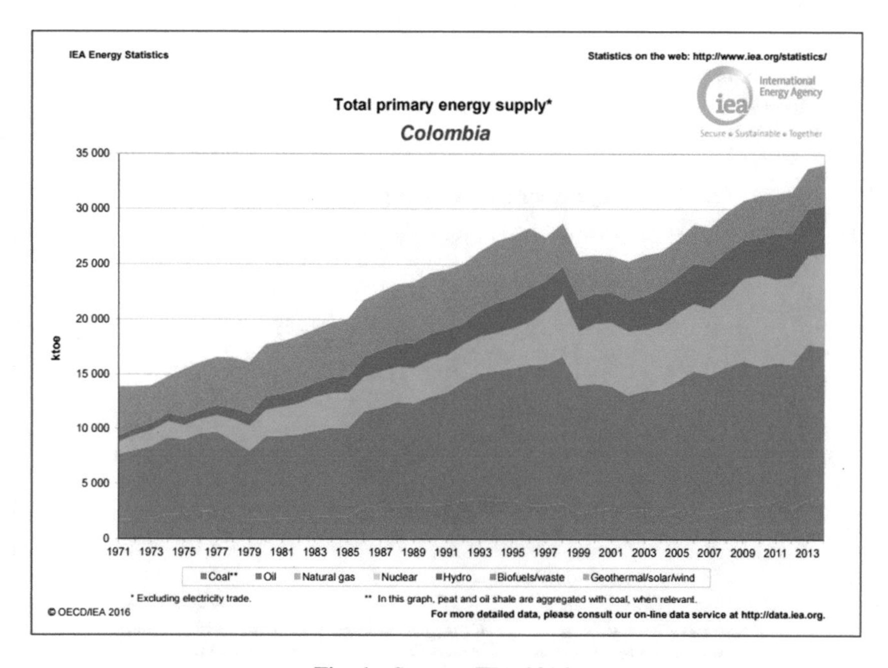

Fig. 1. Source: (IEA 2016)

In Fig. 2, it can be seen that the main energy source, the oil supply, has remained stable throughout this period, exceeding 40% of the Colombian total in 2014. Other fossil sources such as natural gas (25%) and coal (11%) are also important sources for covering the supply. The energy supply provided by renewable sources (hydroelectric, biomass, and waste) covers 23% of the total.

For the same year, as it can be seen in Fig. 3, the demand generated by the internal energy consumption, reached 25,635 ktoe (1.07 EJ). The most frequent user was the transportation sector (30% with more than 10 thousand ktoe). The second line is occupied by industry (19%, 6,408 ktoe), followed by households (14%, 4,951 ktoe). Energy applications for agricultural use and public service management occupy 6% and 5%, respectively. Other specifications together account for approximately 26%.

[2] Note that the result presented is net, since exports are subtracted from gross production.

However, the level of energy use is limited not only to the supply and its sources but also depends to a large extent on the final use of the resource itself.

As mentioned by Ramírez Triana "For example, the light used to read this text is the result of the transformation of electrical energy into light, which in turn was preceded by the transformation of potential energy—accumulated in a reservoir—into electrical energy, released through a hydroelectric turbine" (Ramírez Triana 2010). Some energy experts such as Dale state that energy can be applied in 3 major action fronts, namely, heating, mobility, and electricity. (Dale 2007).

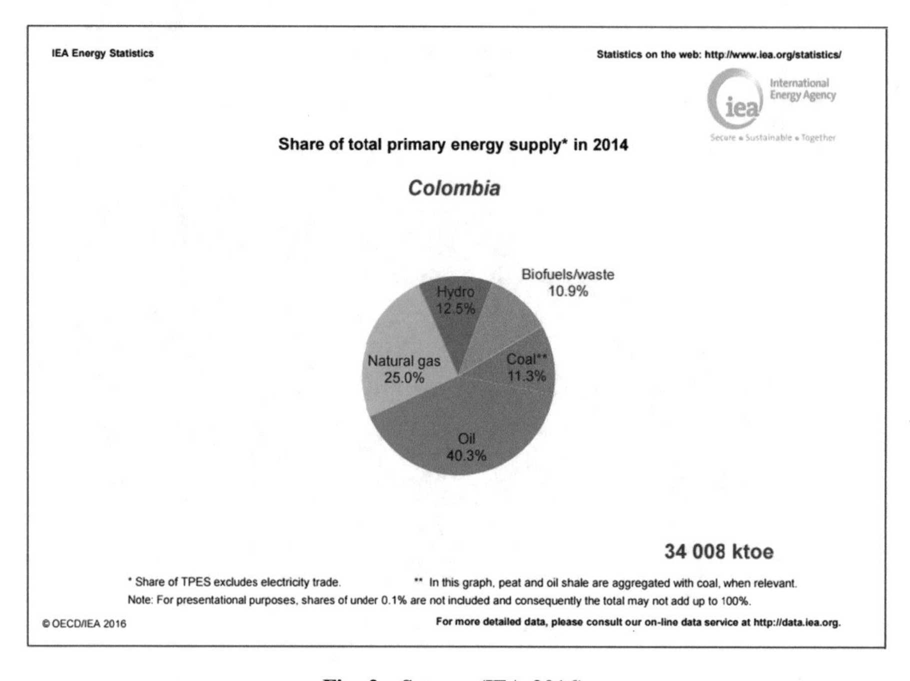

Fig. 2. Source: (IEA 2016)

The provision of heat in homes, for housing, and comfort purposes, is an explicit need in non-tropical places; however, in Colombia, heat is used mainly for cooking. In the industry, it is used for the generation of power and/or electricity from generators under the Faraday–Lenz principle. These heat generators can be fed directly by mineral coal and/or waste or biomass sources. Likewise, natural gas is frequently used to supply this need.

Some marginal solar thermal energy applications are used for purposes such as water heating and bio-acclimatization on a smaller scale. In addition, other initiatives such as concentrating solar energy are now starting to be commercially available in the country. There are also recent advances in the use of biogas through biodigesters wherein the product can be used directly for heating or indirectly for the generation of electricity.

In the case of mobility, Colombia is heavily supported by gasoline and diesel. Under current regulations, these 2 types of fuels are processed in a mixture with cane ethanol and palm biodiesel in combinations of at least 5% biofuel and the rest of fossil origin (Ramírez-Triana 2012). The automotive fleet powered by electric power is still too incipient to be reflected in National Energy Indicators.

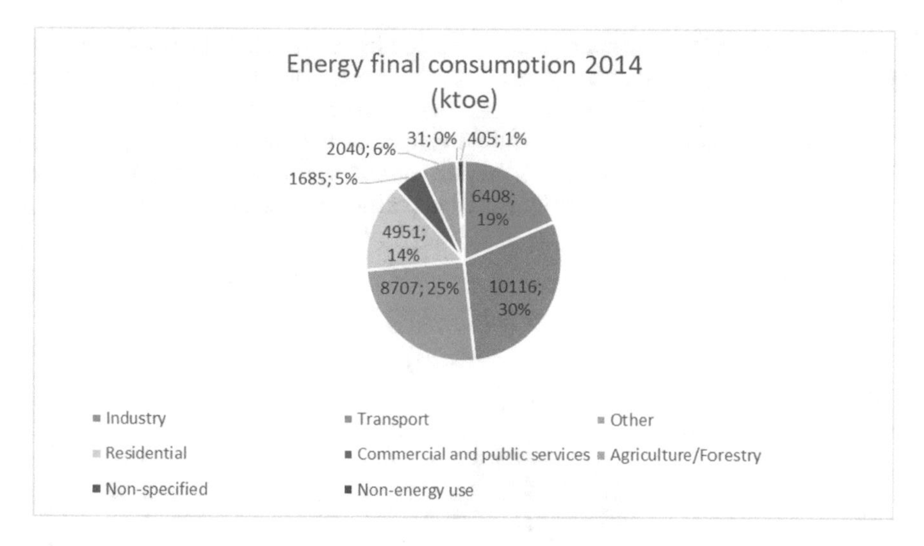

Fig. 3. Prepared by the Authors. Data source: (IEA 2016)

Finally, electricity requires a separate analysis. It is the most versatile type of energy, because it can be applied in all the devices (industrial or residential) that can be connected to an electrical outlet.

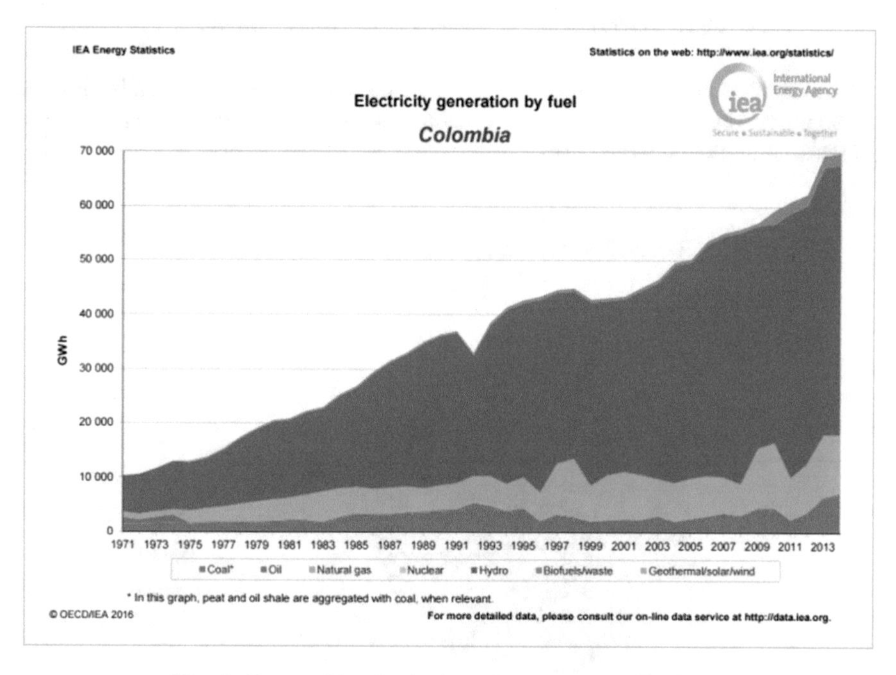

Fig. 4. Prepared by the Authors. Data source: (IEA 2016)

For several decades, the main source of electricity in the national territory has been hydroelectric projects. Other secondary sources of electricity are supported by fossil resources (coal and natural gas), leveraging thermogenerators, as mentioned above (see Figs. 4 and 5). It is important to highlight that renewable alternatives, supported by biomass, are gradually gaining space in the provision of electricity through co-generation systems, whereas oil-based electricity has been practically obsolete since the mid-70 s.

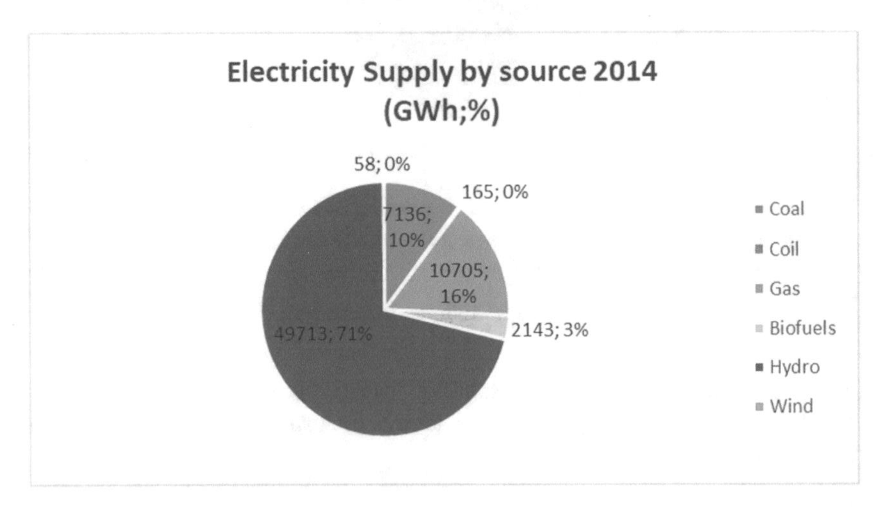

Fig. 5. Prepared by the Authors. Data source: (IEA 2016)

Now, the use given to electricity is summarized in Fig. 6, where it can be observed that the provision of this type of energy mainly provides to residential needs, followed by industrial and commercial uses.

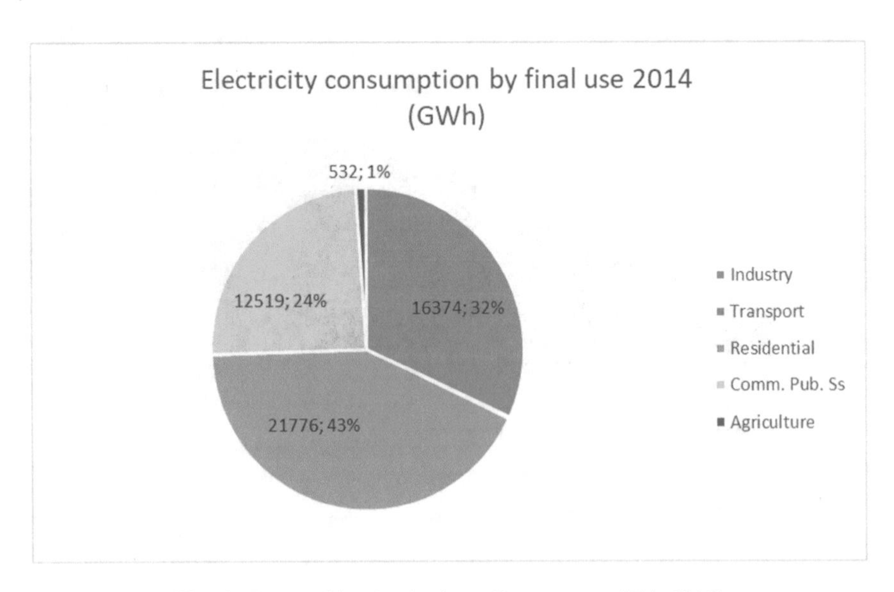

Fig. 6. Prepared by the Authors. Data source: (IEA 2016)

To a lesser extent, agricultural requirements are met. It is precisely this fact that is paramount for better understanding the importance of electricity supply, because households are the main recipients of this type of energy. Given the above, identifying the coverage of the National Electricity System becomes critical, as discussed below.

2.1 Electric Coverage in Colombia

In Colombia, the Colombian Mining and Energy Information System (SIMEC), through the Colombian Electricity Information System (SIEL), is responsible for collecting information regarding the provision of electricity throughout the national territory.

The Colombian reality in terms of the supply of electricity is good, since the Electric Power Coverage Index (ICEE) is close to 97% of the population. This is subdivided into coverage of 99.72% of the main urban areas (known as municipal capitals) and of 87.83% of the rest of the territory wherein the population density is lower (SIEL 2016).

At a quick glance, the last 2 bars in Fig. 7 prove that 22 out of the 33 departments of the country have at least 90% coverage, and in 6 others, coverage reaches at least 80%. Therefore, the national situation is considered positive.

However, there are some parts of the country that face important difficulties in terms of infrastructure and access to the national grid. For the year 2015 the most neglected departments (coverage below 75%) were La Guajira, Vaupés, Amazonas, Putumayo, and Vichada. Figure 8 presents an overview of these problems.

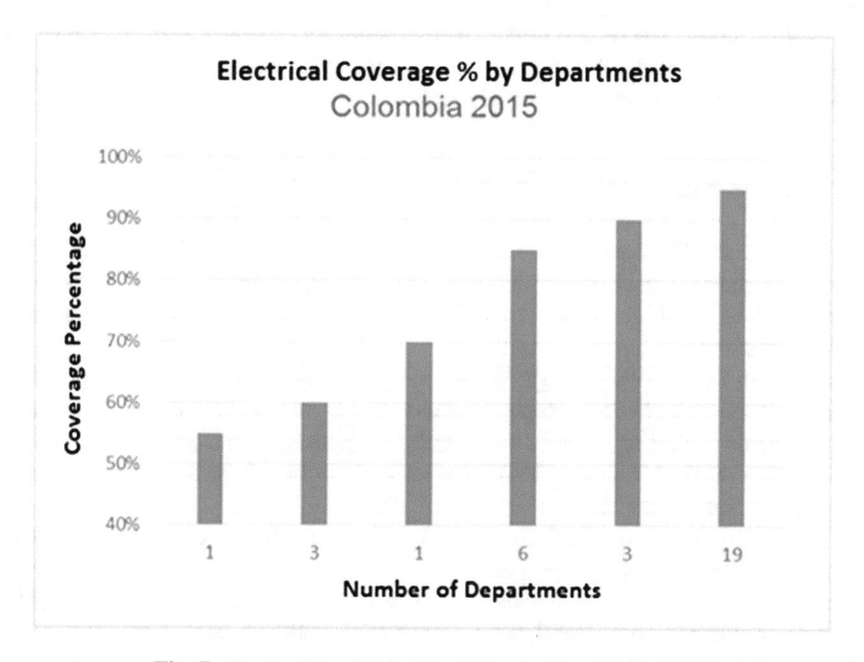

Fig. 7. Prepared by the Authors. Data source: (SIEL 2016)

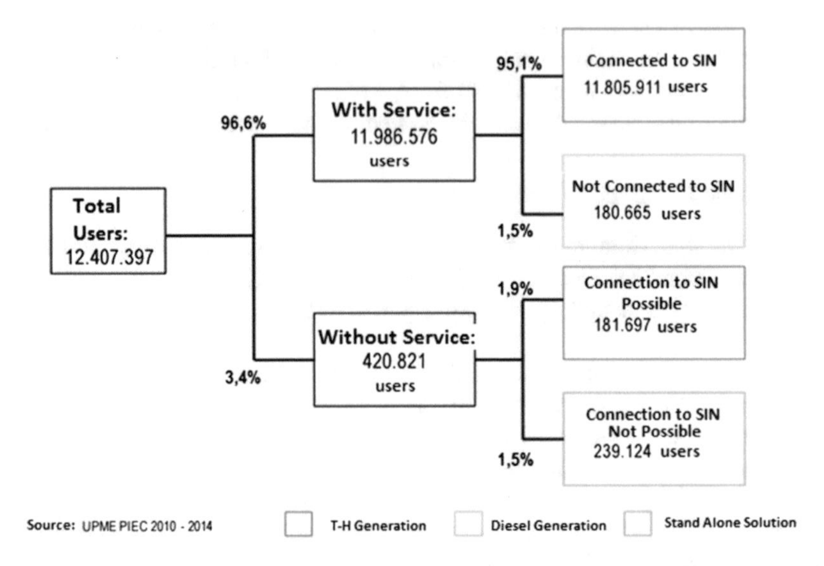

Fig. 8. Users of the national electrical grid in Colombia. Source (UPME 2012).

Non-Interconnected Zones (ZNI) are understood as municipalities, districts, localities, and settlements not connected to the National Interconnected System (SIN), (UPME 2012). Figure 8 describes the population distribution by number of users. The first group is subdivided into users covered under the national grid and people with their own means of supply (not connected to the SIN), basically using power plants powered by diesel. The second group, those without service, are classified as potential users (that is, they can be connected to the SIN) and those that cannot be covered by the system and must seek another type of solution.

3 The Forgotten Effects Model

As proposed by Kaufmann and Gil-Aluja (1988), the Theory of Forgotten Effects provides a mathematical model that uses the notions of fuzzy sets and fuzzy relationships to establish cause–effect relationships that go unnoticed in almost all phenomena or events.

Based on the concepts of fuzzy relationships and fuzzy sets (Gil-Lafuente 2015), the Forgotten Effects Model accepts the notion that there could often be the possibility of ignoring or neglecting some of these causal relationships, and, therefore, provides a mathematical basis for finding those forgotten first- and second-order relationships (Vizuete Luciano et al. 2013).

The Theory of Forgotten Effects is based on the use of fuzzy semantic judgments to describe incidence relationships (Gil-Lafuente 2015). This is achieved through the participation of subject matter experts who assess the veracity of the incidence according to its degrees of intensity (Rodríguez Rubinos et al. 2008). A subject matter expert is generally defined as an individual with vast experience in an area of knowledge or determined activity who contributes with their knowledge and expertise to the research by autonomously evaluating each level of incidence (Salazar-Garza 2012). As described by Gento et al. (2001) The forgotten effects model may be used for analysis of all types of decision making processes, with particular suitability for application in the fields of management and economics.

In this research, the FuzzyLog software was used. It is a calculation program specially designed by its developers to work under the Forgotten Effects Model. This program displays second-order incidence relationships and solves the calculation of the incidence matrices, yielding the results of forgotten effects in graphic and numerical form for analysis by the research team.

4 Implementation of the Forgotten Effects Model to Study ZNI in Colombia

In accordance with the research development stages under the Forgotten Effects Model, in the first instance, the causes and consequences existing around the problem of the lack of connectivity to the electrical grid in Colombian ZNIs were identified. This identification was made based on an extensive review of the literature individually conducted by members of the research team. Subsequently, and based on the findings of this literature review, the entire team of researchers met in a brainstorming session to analyze and decide via consensus the causes and consequences that would form the basis of the research instrument. As a result of this process, as detailed in Table 1, 13 central causes were identified for the ZNI problem in Colombia, and 7 consequences or effects can result from said problem.

With the two subsets (causes and consequences) identified, the level of incidence was measured to find incidences of first and second orders (Gil Lafuente and Barcellos De Paula 2010). For this, we sought to establish a degree of incidence by means of a quantitative value on a Likert scale (Between 1 and 10), and a research instrument matrix was designed, with the following matrices: 1. Causes–Consequences Matrix; 2. Causes–Causes Matrix; and 3. Consequences–Consequences Matrix.

As mentioned above, the Forgotten Effects Model requires consultation with experts who can provide an informed opinion on the subject in question. The research team identified experts from the public, private, and academic sectors, who had a direct relationship with the study, regulation, generation, or distribution of electricity in the country. A formal invitation to participate in the research was distributed, receiving from each expert an answer accepting, declining, or delegating their participation in the study. The invitation was accepted or delegated by 10 experts, 7 of whom took an effective part in the investigation, completing the instrument in its entirety and delivering it to the research team for processing and analysis. Table 2 lists the entities that participated in the study, identifying the sector to which they belong. For reasons of confidentiality the name of the participating experts is not reported.

Table 1. Causes and consequences of the lack of connectivity to the electrical grid in Colombian ZNIs

Causes		Consequences	
1	Costs	1	Other industries or sectors feel discourage to invest in the ZNI
2	Firms ineffectiveness	2	Isolation of community (from general media, development, education, etc.)
3	Public tender process to award projects in ZNI	3	Restraint to competitiveness
4	Corruption	4	Local inflationary effects
5	Poverty	5	Escalation of the armed conflict
6	Geophysics-Natural conditions	6	Opportunities for development of local technological advances
7	Presence of armed violent groups	7	Environmental degradation
8	Risk to contractor		
9	Level of private-public attention to the ZNI		
10	Low financial interest in the ZNI		
11	Local technological advances		
12	Poor infrastructure to access to ZNI		
13	Access to funds		

Source: Prepared by the Authors.

Table 2. Organizations participating in the research

Acronym	Entity	Sector
CECODES	Colombian Business Council for Sustainable Development	Private
MME[a]	Ministry of Mines and Energy	Public
SER	Association of Renewable Energies, Colombia	Private
OCE	Colombian Energy Observatory. National University of Colombia	Academia
UPME	Energy Mining Planning Unit	Public
POLI	Institución Universitaria Politécnico Grancolombiano	Academia

[a]Two experts from the Ministry of Mines and Energy participated.
Source: Prepared by the Authors

For the aggregation of the answers received, in this study, we used the Weighted Aggregation on Fuzzy Logic method - WAFL (Ramírez-Triana et al., Forthcoming) designed by the research team as a weighted aggregation method that allows for the implementation of the Forgotten Effects Model taking into account the weighted variables that accurately reflect the opinion of each expert, and supports reaching a unified response among the specialists in the group. For the weighting of the answers received, the following information was requested from the participating experts:

1. Educational Level (options: Undergraduate, Graduate, Master's, PhD)
2. Type of Organization (options: Public, Private, and Academic)
3. Years of experience in the sector
4. Self-assessment of their Level of Experience in the Energy Sector*
5. Self-assessment of their Level of Experience in the Renewable Energy Sector*
6. Self-assessment of their Level of Experience in the ZNI Sector*

*Self-assessment was conducted on a scale of 1–5, where 5 was vast experience and 1 little experience

The WAFL method was used to minimize heterogeneity in the responses from the experts, and obtain a single weighted result representative of the opinion of the participating experts. This weighted result was used by the research team to process additional data using the FuzzyLog software and to identify the forgotten effects (Gil Lafuente and Bassa 2011). Tables 3, 4 and 5 below show the matrix results that were used as input for the software and the identification of the first and second-order relationships.

Table 3. Causes–Consequences matrix weighted by the WAFL method

Causes–Consequences	E1	E2	E3	E4	E5	E6	E7
C1	0,7	0,7	0,7	0,6	0,4	0,5	0,5
C2	0,6	0,5	0,5	0,4	0,3	0,5	0,5
C3	0,6	0,5	0,5	0,3	0,3	0,5	0,3
C4	0,8	0,7	0,7	0,6	0,5	0,4	0,6
C5	0,6	0,6	0,6	0,4	0,7	0,5	0,7
C6	0,6	0,7	0,7	0,6	0,6	0,6	0,4
C7	0,9	0,8	0,8	0,6	0,9	0,5	0,7
C8	0,8	0,6	0,7	0,4	0,4	0,4	0,4
C9	0,7	0,7	0,6	0,5	0,5	0,5	0,5

Table 4. Causes–Causes matrix weighted by the WAFL method

Cause–Cause	C1	C2	C3	C4	C5	C6	C7	C8	C9	C10	C11	C12	C13
C1	1	0,6	0,4	0,4	0,5	0,2	0,3	0,5	0,6	0,7	0,6	0,8	0,5
C2	0,7	1	0,4	0,5	0,5	0,2	0,3	0,4	0,5	0,4	0,5	0,6	0,4
C3	0,5	0,5	1	0,7	0,5	0,2	0,3	0,4	0,6	0,5	0,4	0,5	0,4
C4	0,8	0,7	0,6	1	0,7	0,1	0,5	0,6	0,5	0,6	0,4	0,6	0,3
C5	0,4	0,2	0,2	0,4	1	0,2	0,5	0,4	0,4	0,6	0,6	0,5	0,5
C6	0,8	0,4	0,2	0,2	0,5	1	0,4	0,6	0,6	0,6	0,6	0,7	0,3
C7	0,7	0,5	0,4	0,7	0,7	0,1	1	0,9	0,8	0,7	0,4	0,7	0,6
C8	0,7	0,4	0,4	0,3	0,3	0,2	0,1	1	0,5	0,5	0,5	0,6	0,4
C9	0,6	0,3	0,5	0,4	0,5	0,2	0,4	0,4	1	0,5	0,5	0,7	0,3
C10	0,6	0,4	0,3	0,3	0,6	0,1	0,2	0,4	0,6	1	0,4	0,5	0,7
C11	0,5	0,2	0,1	0,2	0,5	0,1	0,1	0,2	0,4	0,2	1	0,5	0,2
C12	0,7	0,5	0,4	0,4	0,7	0,2	0,4	0,7	0,6	0,7	0,5	1	0,5
C13	0,6	0,4	0,4	0,4	0,5	0,1	0,2	0,5	0,5	0,4	0,5	0,4	1

Source: Prepared by the Authors

Table 5. Consequences–Consequences matrix weighted by the WAFL method

Consec–Consec	E1	E2	E3	E4	E5	E6	E7
E1	1	0,7	0,7	0,6	0,5	0,5	0,5
E2	0,8	1	0,7	0,6	0,5	0,5	0,6
E3	0,6	0,5	1	0,7	0,4	0,3	0,3
E4	0,4	0,3	0,6	1	0,3	0,4	0,4
E5	0,9	0,8	0,8	0,7	1	0,4	0,7
E6	0,2	0,1	0,1	0,2	0,1	1	0,1
E7	0,5	0,5	0,5	0,5	0,3	0,4	1

Source: Prepared by the Authors

The feeding of the FuzzyLog software with the information from the three previous matrices yielded the necessary calculations for the identification of the direct and indirect causal relationships accumulated through the matrix of second-order incidences, as shown in Fig. 9, and the matrix of Forgotten Effects, as shown in Fig. 10.

	E_1	E_2	E_3	E_4	E_5	E_6	E_7
C_1	0,8	0,7	0,7	0,7	0,7	0,6	0,7
C_2	0,7	0,7	0,7	0,7	0,5	0,5	0,6
C_3	0,6	0,6	0,6	0,6	0,5	0,5	0,6
C_4	0,8	0,7	0,7	0,7	0,7	0,5	0,7
C_5	0,7	0,7	0,7	0,7	0,7	0,6	0,7
C_6	0,7	0,7	0,7	0,7	0,6	0,6	0,6
C_7	0,9	0,8	0,8	0,7	0,9	0,5	0,7
C_8	0,9	0,8	0,8	0,7	0,9	0,6	0,7
C_9	0,8	0,8	0,8	0,7	0,8	0,6	0,7
C_{10}	0,8	0,7	0,7	0,7	0,7	0,6	0,7
C_{11}	0,6	0,6	0,6	0,6	0,6	0,6	0,6
C_{12}	0,8	0,8	0,8	0,7	0,7	0,6	0,7
C_{13}	0,7	0,7	0,7	0,7	0,6	0,6	0,6

Fig. 9. Matrix of second-order effects. Source: FuzzyLog

	E_1	E_2	E_3	E_4	E_5	E_6	E_7
C_1	0,1	0	0	0,1	0,3	0,1	0,2
C_2	0,1	0,2	0,2	0,3	0,2	0	0,1
C_3	0	0,1	0,1	0,3	0,2	0	0,3
C_4	0	0	0	0,1	0,2	0,1	0,1
C_5	0,1	0,1	0,1	0,3	0	0,1	0
C_6	0,1	0	0	0,1	0	0	0,2
C_7	0	0	0	0,1	0	0	0
C_8	0,1	0,2	0,1	0,3	0,5	0,2	0,3
C_9	0,1	0,1	0,2	0,2	0,3	0,1	0,2
C_{10}	0	0	0	0,2	0,3	0,1	0,4
C_{11}	0,3	0,1	0	0,2	0,4	0	0,2
C_{12}	0	0	0	0	0,1	0	0,1
C_{13}	0,2	0,2	0	0,3	0,3	0	0,3

Fig. 10. Matrix of forgotten effects. Source: FuzzyLog

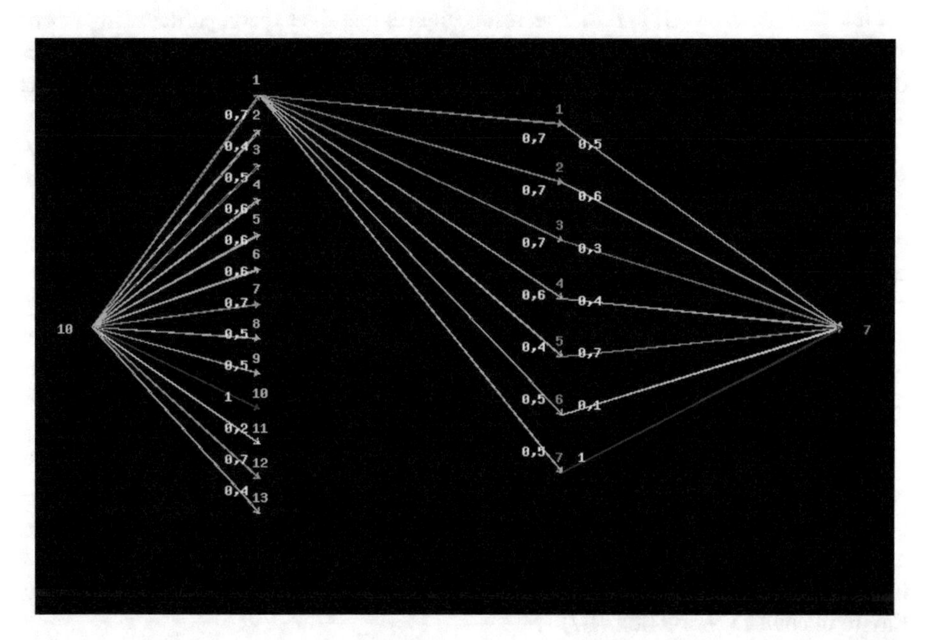

Fig. 11. Total Incidence of low financial interest toward the ZNI cause on the environmental deterioration effect. Source: FuzzyLog

In particular, it is interesting to analyze the forgotten effect of 0.4, identified as low financial interest toward ZNIs (C10) on environmental deterioration (E7), as shown in Fig. 11. This effect would imply that the environmental deterioration in the ZNIs is indirectly caused by low financial interest in the area, which creates an environment conducive to the presence of armed groups, with direct involvement in environmental degradation events, such as deforestation, water, and atmospheric pollution.

5 Discussion and Conclusions

This research study proposes the analysis of the causes for the lack of technological development in Non-Interconnected Zones (ZNI) to the National Electrical Grid in Colombia, and their corresponding consequences. The bridge that connects the elucidation of this problem in terms of its causes and consequences is materialized through the application of the Theory of Forgotten Effects based on fuzzy logic. To gather information, a research instrument was designed and implemented with subject matter experts in energy and ZNIs, which allowed for the collection and analysis of first- and second-order relationships in incidence matrices.

When analyzing the first-order incidents identified in the original matrix (Table 3), we must focus our attention on the strongest and weakest relationships identified. As for the relations of greater intensity, it is observed, in the first case (C7, E1), that presence of armed groups is a strong disincentive to general investment in the ZNIs. Also, in the second relationship (C7, E5), as expected, there is a strong incidence of the presence of armed groups on conflict escalation.

However, when analyzing the first-order incidents of lower intensity, a relation of 0.2 is found between (C11, E5). This result implies that the experts participating in this study do not see a strong incidence of the existence of local technological advances (that may totally or partially solve the energy needs in ZNIs), in the escalation of the armed conflict.

The analysis of the Matrix of Forgotten Effects (Fig. 10) identifies an effect of 0.4 that had been overlooked in the relationship between C11, E5. This forgotten effect indicates that local technological advances explain, to a certain degree, the escalation of the armed conflict. A possible explanation would be that armed groups are not interested in the progress of the affected communities, and, therefore, they allow the conflict to increase. Another probable approach is that experts have assumed only the magnitude of the relationship, but not the direction. For example, they may consider that technological advances have influence on the conflict, but this influence is on its reduction.

Likewise, the results obtained with the Forgotten Effects approach lead to the identification of 32 of the 91 relationships in which the forgotten effect is equal to zero; which is equivalent to 35.16% of causal relationships in which all the effects have been duly captured and would not present incidences that have been ignored and may have an impact on the problem of ZNIs. The research progress reported in this paper highlights, in particular the effect of Low Financial Interest Towards ZNI (C10) on Environmental Deterioration (E7).

Highlighting the importance of adopting light innovation as a power supply tool for these areas, the research has focused on the difficulty of providing Colombian ZNIs with efficient and continuous electrical power services, and the relationship between said difficulty and corruption and lack of access roads, drinking water, and health services, among others. The research team continues analyzing the information gathered, which will allow the team to approach the problem from a sectoral perspective, among others, and provide other additional results that may be useful for the academic community and for the designers and implementers of public policy to contribute to the positive management of non-interconnected areas in the country.

Acknowledgements. This paper is part of a research project supported by the "Red Iberoamericana para la Competitividad, Innovación y Desarrollo" (REDCID) identified with number 616RT0515 and part of the "Programa Iberoamericano de Ciencia y Tecnología para el Desarrollo" (CYTED). The authors would like to thank the following institutions for their valuable participation in this research: Consejo Empresarial Colombiano para el Desarrollo Sostenible – CECODES, Ministerio de Minas y Energía, Asociación de energías renovables Colombia – SER Colombia, Observatorio Colombiano de Energía Universidad Nacional de Colombia – OCE, Unidad de Planeación Minero Energética – UPME, and Institución Universitaria Politécnico Grancolombiano.

References

Dale, B.: Thinking clearly about biofuels: ending the irrelevant 'net energy' debate and developing better performance metrics for alternative fuels. Biofuels Bioprod. Biorefin 14–17 (2007)

Gento, A., Lazzari, L.L., Machado, E.A.M.: Reflexiones acerca de las matrices de incidencia y la recuperación de efectos olvidados. Cuadernos Del CIMBAGE **4**, 11–27 (2001)

Gil-Lafuente, A.M.: Application of the forgotten effects model to the economic effects for public European health systems by the early diagnostics of emergent and rare diseases. Procedia Econ. Finan. **22**, 10–19 (2015). https://doi.org/10.1016/S2212-5671(15)00221-X

Gil-Lafuente, A.M., Barcellos De Paula, L.: Una Aplicación De La Metodología De Los Efectos Olvidados: Los Factores Que Contribuyen Al Crecimiento Sostenible De La Empresa. Cuadernos Del CIMBAGE **12**(12), 23–52 (2010)

Gil Lafuente, A.M., Luis Bassa, C.: Identificación de los atributos contemplados por los clientes en una estrategia CRM utilizando el modelo de efectos olvidados. Cuadernos Del CIMBAGE **13**(13), 107–127 (2011)

IEA. Statistics on Colombia: Indicators for 2014 (2016). https://www.iea.org/statistics/?country=COLOMBIA&year=2016&category=Key%20indicators&indicator=TPESbySource&mode=chart&categoryBrowse=true&dataTable=BALANCES&showDataTable=false). Accessed 16 Feb 2018

Kaufmann, A., Gil Aluja, J.: Modelos para la investigación de efectos olvidados. Editorial Milladoiro, Vigo (1988)

OECD. Estudios Económicos de la OCDE Colombia. Evaluación Económica Paris (2013)

Pineda-Escobar, M.A.: Urban agriculture as a strategy for addressing food insecurity of BoP populations. In: Casado-Cañeque, F., Hart S.L. (ed.) Base of the Pyramid 3.0: Sustainable Development through Innovation and Entrepreneurship. Greenleaf Publishing Limited, Sheffield (2015)

Ramírez Triana, C.A.: Biocombustibles: seguridad energética y sostenibilidad. Conceptualización académica e implementación en Colombia. Punto de Vista, 45–70 (2010)

Ramirez Triana, C.A.: Establecimiento De Una Industria Bioenergética Sostenible En Colombia. Cap&Cua, 1–15 (2012)

Rodríguez Rubinos, J.M., Ramírez Reyes, M.A., Díaz Pontones, V.: Efectos olvidados en las relaciones de causalidad de las acciones del sistema de capacitación en las organizaciones empresariales. Revista de Métodos Cuantitativos Para La Economía y La Empresa **5**, 29–48 (2008)

Salazar-Garza, R.: El peso mexicano: la gestión de cobertura del riesgo cambiario mediante la Teoría de los Efectos Olvidados. J. Econ. Finan. Adm. Sci. **17**(32), 53–73 (2012)

SIEL. Cobertura de Energía Eléctrica a 2015 (2016). Retrieved from Consulta de estadísticas: http://www.siel.gov.co/Inicio/CoberturadelSistemaIntercontecadoNacional/ConsultasEstadisticas/tabid/81/Default.aspx. Accessed 16 Feb 2018

UNDP Human Development Report 2016. Human Development for Everyone. United Nations Development Programme, New York (2014)

UPME. Acciones y retos para energización de las ZNI en el país. Ministerio de Minas y Energía, Bogotá (2012)

Vizuete Luciano, E., Gil-Lafuente, A.M., García González, A., Boria-Reverter, S.: Forgotten effects of corporate social and environmental responsibility. Kybernetes **42**(5), 736–753 (2013). https://doi.org/10.1108/K-04-2013-0065

Innovation Capabilities and Innovation Systems: A Forgotten Effects Analysis of Their Components

Gerardo G. Alfaro-Calderón[1], Artemisa Zaragoza[1], Víctor G. Alfaro-García[2](✉), and Anna M. Gil-Lafuente[3]

[1] Facultad de Contabilidad y Ciencias Administrativas, Universidad Michoacana de San Nicolás de Hidalgo, Morelia, Mexico
ggalfaroc@gmail.com, artemisazaragoza@hotmail.com
[2] Facultad de Economía, Universidad Autónoma de San Luis Potosí, San Luis Potosí, Mexico
valfaro06@gmail.com
[3] Facultat d'Economia y Empresa, Universitat de Barcelona, Barcelona, Spain
amgil@ub.edu

Abstract. An innovation system can be considered an entity in charge of coordinating all the actors involved in the innovation processes of a country or region. Recently, the Mexican innovation system has adopted a model to assess the behavior and measurement of the resources invested in innovation activities. The innovation model comprises seven areas denominated as innovation pillars. Applying the forgotten effects theory, this paper aims to quantify the effects of the results of the innovation pillars on the innovation capabilities of a region. The results show a general direct effect of all the pillars on innovation capabilities; however, the pillars' institutions, infrastructure and market sophistication present significant multiplicative indirect effects that suggest that the initial effects should be reviewed.

Keywords: Forgotten effects theory · Innovation system · Innovation capabilities · Regional development

1 Introduction

The purpose of the National Science, Technology and Innovation System (NSTIS) is to enhance the coordination and cooperation of activities related to research and development in science, technology and innovation in Mexico. The NSTIS is made up of institutions from three interlocking sectors: public, academic, and research and business.

The NSTIS affects two different environments. First, it affects the political environment, namely, through the General Council for Scientific Research, Technological Development and Innovation, the Special Program for Science, Technology and Innovation (PECITI, Programa Especial de Ciencia, Tecnología e Innovación), and the regional and sectorial programs. Second, it influences the legal environment, which includes all the managerial, economic and legal factors related to the country's normativity that contribute to scientific innovation and technological development. It also

J. C. Ferrer-Comalat et al. (Eds.): MS-18 2018, AISC 894, pp. 51–62, 2020.
https://doi.org/10.1007/978-3-030-15413-4_5

includes the public administration of dependencies and entities, institutions from both the public and private sectors, and state and federal governments within its concentration, coordination and linkages. In this group, we can also place the national researchers' system and research networks.

The NSTIS in Mexico is coordinated by the head of the National Council for Science and Technology (CONACYT). According to the Science and Technology Law, the NSTIS's purpose is to enhance the scientific and technological innovation capacity by focusing on solving problems that have been classified as priorities for the development and wellness of the population. Some of the goals of the NSTIS are, for example,

- Promoting the development and linkage of basic science, technological development and innovation mainly by focusing on updating and improving quality on education and expanding the frontiers of knowledge to create a knowledge-based society.
- Technology transfer as a driver of growth, transforming productive processes and services with the purpose of making the national productive sector an entity with a world-class level of productivity and competitiveness.
- Strengthening the regional development through comprehensive decentralization policies that foster scientific, technological and innovation activities.
- Promoting gender equality in all system areas.

One of the key goals for this research is aligned with the purpose of the NSTIS to facilitate processes that allow the definition of priorities of federal government resources for science, technology and innovation, as well as the allocation and optimization of those resources. However, according to current information on innovation, the National Innovation System objectives have not been fully achieved; that is, previous works suggest that human capital development and business sophistication are the two areas considered in the NSTIS that need more investment to reach a higher rank on innovation indexes; nonetheless, infrastructure is the area which currently receives more investment.

Mexico's innovation index is obtained through the use of a generalized innovation model to measure innovation impact. This model was jointly developed by the Mexican Ministry of Economy and the Venture Institute in 2013. It is comprised of five input pillars: institutions, infrastructure, human capital and research, market sophistication, and business sophistication. The model also considers two resulting output pillars: knowledge and technology and creative products and services.

This method allows for the identification and characterization of the key factors inherent in innovation. It also offers a particular view of a regional economic and innovative activity (Comité Intersectorial para la Innovación 2011; Venture Institute 2013).

The purpose of this paper is to find the impact of forgotten effects of innovation pillars on the innovation capabilities of enterprises. By using this methodology, it can be observed which areas of the NSTIS are driving which innovation capabilities in a region. This measurement is especially important because it allows for the visibility of areas that could exert a higher multiplicative effect on a regions' innovation capabilities, which might not initially be considered.

The paper reminder is as follows. Section 2 provides the theoretical background of the study. Section 3 presents the methodology of the forgotten effects theory. Section 4 summarizes our main findings, while Sect. 5 provides a discussion of those findings. Finally, Sect. 6 presents the concluding comments.

2 Theoretical Background

2.1 Foundations of the Study

Economic development implies growth and income distribution based on the generation and diffusion of endogenous technological capacities, as well as a structural change related to new products, managerial processes, and production processes.

Products, processes and services developed by firms have constantly improved since the industrial revolution and so have life standards. Currently, innovation has acquired a special relevance in a country's economy due to globalization. Each of these factors have forced all countries to boost their economic activities along the value chain in order to guarantee competitiveness and prosperity for a nation.

Research shows a close relationship between innovation capacities and economic success (Geoffrey 2009), both at the macro- and microeconomic level. At the macroeconomic level, almost 50% of the disparity between a country's per capita income is produced by differences in technological advances and innovation influences. At the microeconomic level, innovative companies show better performance and create more and better jobs in all sectors. According to an innovation analysis made by the Organization for Economic Cooperation and Development (OECD 2014), innovation also increases the labor productivity of firms.

In Latin America, Mexico is one of the highest investors of GDP in R&D, and statistics show that the net investment is approximately 0.49%, which is 58th place on the global innovation ranking. On the other hand, Chile has an expenditure in the same area of 0.036%–, which is the lowest of all the OECD members–, and yet Chile is in the 46th place on the same ranking. Taking innovation models as a point of study and comparison, the Swiss model covers six main innovation areas, while the Mexican model covers five and the Chilean model two. However, both Chile and Switzerland are better placed in the innovation ranking. One particular difference between the countries is that Switzerland and Chile have changed their investment policies, focusing on an open innovation model (OECD 2014; Cornell University 2017).

2.2 Innovation Variables Model

All the variables that shape Mexico's innovation model (Venture Institute 2013) are designated innovation pillars. This model is comprised of five inputs: Institutions, Human Capital and Research, Infrastructure, Market Sophistication, and Business Sophistication. It also includes two outputs: Technology and Knowledge and Creative Goods and Services.

The pillar of institutions valuates the institutional frame in which innovation is developing. A good institutional frame allows for a better environment for entrepreneurs and business development. This pillar is formed by the following indexes: delinquency, governmental efficiency, press liberty and security, quality of regulation, ease of opening a new business and ease payoff paying taxes.

The second pillar, human capital, values research and human capital quality. This pillar is formed by: expenditure on education, education level, PISA test results, education quality, upper level education, masters and PhDs, scholarships, number of national researchers, R&D investment and research institutions.

The third pillar, infrastructure, makes both information and people flow easier; it also helps in the exchange of ideas, services and products. Some indexes that are used to measure this pillar are ICT access, Internet use, e-government, online participation, communication of information, energy production, electrical intake, transportation quality, gross fixed capital formation, energetic efficiency, and environmental performance.

Market sophistication is the fourth pillar. It was designed to measure readiness and credit access. This pillar's indicators are credit access, use of credit, microfinance, market capitalization, stock exchange, and local competition intensity.

Business sophistication, the fifth pillar, values the business environment of a specific region, characterizing the enterprises as a fundamental item for national innovation processes. This pillar is formed by employed professionals, enterprises who offer training, business R&D, R&D business expenditure, strategic alliances, clusters, intellectual property payment, high-tech imports, foreign direct investment, incubators and accelerators, and the quantity of startups and SMEs.

The technology and knowledge pillar measures new regional technology and knowledge production. It values the amount of knowledge and its impact on the economy. This pillar is formed by patents, patent growth rate, published articles, labor productivity, software use, innovation impact on market, intellectual property use, and high-tech exports.

Finally, the creative goods and services pillar values the population's creative capacity by measuring regional cultural access and production. This pillar is formed by ICT and business model creation, ICT and organization model creation, cultural activities expenditure, movie and TV production, newspaper presence, and tweets per capita.

2.3 Innovation Capabilities

Mexico's innovation agenda aims to enhance the regional capabilities of local firms. To quantify the degree of development that any chosen sector could provide in terms of innovation spillover, this research takes the proposal developed by Adams et al. (2006) as the main innovation measurement, retrieved and referenced by Alfaro et al. (2017). Adams et al.'s (2006) proposal is based on the review of six models and frameworks of innovation measurement. In their study, from the seven main areas outlined by the authors, the resulting framework for measuring innovation takes into account recurrent and relevant factors when quantifying the structural capabilities of companies to make and maintain continuous change.

Adams' model is comprised of seven areas that include Innovation Strategy, Knowledge Management, Project Management, Portfolio Management, Internal Drivers, Organization and Structure, and External Drivers.

Innovation Strategy is formed by indicators such as strategic orientation and strategic leadership. Knowledge Management is formed by knowledge repository, idea generation and information flow. Project Management includes project efficiency, communication and collaboration. Portfolio Management balances risk-return and optimal usage of tools. Organization and Structure is made up of culture and structure. Finally, External Drivers are comprised of market research, market testing and marketing and sales (Adams et al. 2006).

3 Methodology

Kaufmann and Aluja (1988) proposed the *theory of the forgotten effects*. This theory allows one to obtain all the direct and indirect relations of the analyzed phenomena with no possibility of errors. According to the authors, all the events that surround us are part of a system or subsystem. Hence, we can almost ensure that any activity is subject to a *cause* and *effect* relationship. However, despite having a good control system, there is always the possibility of leaving behind, either voluntarily or involuntarily, some causal relationships that are not always explicit, obvious, or visible. Therefore, it is not uncommon to find *hidden* effects of the analyzed phenomena due to the incidence of other effects or outcomes. The forgotten effects theory is an innovative and efficient approach that considers all the relations among phenomena, minimizing errors that may occur during modeling (Gil-Lafuente 2005). In this study, we have two sets of elements: causes and effects (Kaufmann and Aluja 1988).

$$A = \{a_i / i = 1, 2, \ldots, n\} : \text{Innovation pillars}$$
$$B = \left\{b_j / j = 1, 2, \ldots, m\right\} : \text{Innovation capabilities}$$

We assume that there is an occurrence of a_i over b_j if the value of the membership functions of the feature pair (a_i, b_j) is estimated at between [0,1] (the value of each cell in the array cannot be smaller than 0 or greater than 1 as if we had valued from 0 to 10 but in decimals), i.e., $\forall(a_i, b_j) \Rightarrow \mu(a_i, b_j) \in [0, 1]$.

The set of valuated elements define the "direct relationship matrix," which shows the cause-effect relations that occur between the joint elements of the *A* set and the joint elements of the *B* set.

$$M = \begin{array}{c|c|c|c|c|c|c|}
 & b_1 & b_2 & b_3 & b_4 & \cdots & b_j \\
a_1 & \mu_{a_1b_1} & \mu_{a_1b_2} & \mu_{a_1b_3} & \mu_{a_1b_4} & \cdots & \mu_{a_1b_j} \\
a_2 & \mu_{a_2b_1} & \mu_{a_2b_2} & \mu_{a_2b_3} & \mu_{a_2b_4} & \cdots & \mu_{a_2b_j} \\
a_3 & \mu_{a_3b_1} & \mu_{a_3b_2} & \mu_{a_3b_3} & \mu_{a_3b_4} & \cdots & \mu_{a_3b_j} \\
a_4 & \mu_{a_4b_1} & \mu_{a_4b_2} & \mu_{a_4b_3} & \mu_{a_4b_4} & \cdots & \mu_{a_4b_j} \\
a_5 & \mu_{a_5b_1} & \mu_{a_5b_2} & \mu_{a_5b_3} & \mu_{a_5b_4} & \cdots & \mu_{a_5b_j} \\
\vdots & \vdots & \vdots & \vdots & \vdots & \vdots & \vdots \\
a_i & \mu_{a_ib_1} & \mu_{a_ib_2} & \mu_{a_ib_3} & \mu_{a_ib_4} & \cdots & \mu_{a_ib_j}
\end{array}$$

To continue with the explanation of the model, below we show how the effects can accumulate. For this, we need to define a third set of elements, which are different from the previous two:

$$C = \{c_k/k = 1, 2, \ldots, z\}$$

This consists of the elements that are the effects of the B set:

$$N = \begin{array}{c|c|c|c|c|}
 & c_1 & c_2 & \cdots & c_z \\
b_2 & \mu_{b_1c_1} & \mu_{b_1c_2} & \cdots & \mu_{b_1c_z} \\
b_2 & \mu_{b_2c_1} & \mu_{b_2c_2} & \cdots & \mu_{b_2c_z} \\
\vdots & \cdots & \cdots & \cdots & \cdots \\
b_m & \mu_{b_mc_1} & \mu_{b_mc_2} & \cdots & \mu_{b_mc_z}
\end{array}$$

and having the common elements of the B set:

$$M = \begin{array}{c|c|c|c|c|}
 & b_1 & b_2 & \cdots & b_m \\
a_1 & \mu_{a_1b_1} & \mu_{a_1b_2} & \cdots & \mu_{a_1b_m} \\
a_2 & \mu_{a_2b_1} & \mu_{a_2b_2} & \cdots & \mu_{a_2b_m} \\
\vdots & \cdots & \cdots & \cdots & \cdots \\
a_n & \mu_{a_nb_1} & \mu_{b_mc_2} & \cdots & \mu_{b_mc_z}
\end{array}
\qquad
N = \begin{array}{c|c|c|c|c|}
 & c_1 & c_2 & \cdots & c_z \\
b_2 & \mu_{b_1c_1} & \mu_{b_1c_2} & \cdots & \mu_{b_1c_z} \\
b_2 & \mu_{b_2c_1} & \mu_{b_2c_2} & \cdots & \mu_{b_2c_z} \\
\vdots & \cdots & \cdots & \cdots & \cdots \\
b_m & \mu_{b_mc_1} & \mu_{b_mc_2} & \cdots & \mu_{b_mc_z}
\end{array}$$

We can therefore say that the P matrix defines the causal relationships between the first set of elements A and the elements of the third C set, with the intensity or degree that leads us to consider the elements belonging to set B.

$$M \subset A \times B \text{ i } \mathrm{N} \subset B \times C.$$

The mathematical operation used to determine the effect of A on C is the max-min composition. In fact, when there are three impact relationships, we would have:

$$M \subset A \times B, \mathrm{N} \subset B \times C, \mathrm{P} \subset A \times C$$

Additionally, the Kaufmann (1977) equation is induced to:

$$M \circ N = P$$

where the $\circ$ symbol represents precisely the max-min composition. The composition of two relations is uncertain such that:

$$\forall (a_i, c_z) \in A \times C : \mu(a_i, c_z)_{\mathrm{M} \circ \mathrm{N}} = V_{bj}\left(\mu_{\mathrm{M}}(a_i, b_j) \wedge \mu_{\mathrm{N}}(b_j, c_z)\right)$$

We, therefore, affirm that the relationship of impact P defines the causal relationships between the elements of the A set and the elements of the C set in intensity or degree, implying which elements belong to the B set.

4 Application of the Model and Results

As mentioned in the previous section, the chosen *causes* are the innovation pillars defined by the innovation variables model. Table 1 gathers the information of the causes. The *effects* are the seven innovation areas first proposed by Adams et al. (2006) and retrieved by Alfaro et al. (2017). Table 2 presents the effects considered in the model.

Table 1. Causes - innovation pillars

Name	Acronym
Institutions	IN
Human capital and research	HCR
Infrastructure	INF
Market sophistication	MS
Business sophistication	BS

Table 2. Effects - innovation capabilities

Name	Acronym
Innovation strategy	IS
Knowledge management	KM
Project management	PrM
Portfolio management	PoM
Internal drivers	ID
Organization and structure	OE
External drivers	ED

We start 3 by valuating an occurrence of a_i over b_j with a_i membership functions estimated between [0,1]. Table 3 shows the valuations.

Table 3. Direct cause-effect matrix

	IS	KM	PrM	PoM	ID	OE	ED
IN	0,5	0,2	0,4	0,5	0,1	0,6	0,7
HCR	0,7	0,6	0,6	0,7	0,8	0,5	0,8
INF	0,5	0,5	0,4	0,3	0,4	0,3	0,5
MS	0,6	0,6	0,5	0,6	0,5	0,4	0,7
BS	0,7	0,6	0,5	0,7	0,4	0,5	0,7

As mentioned in the previous sections, the direct effects among phenomena are not enough to make an in-depth analysis given that causes (innovation pillars) are conditioned by themselves, and effects (innovation capabilities) are affected not only by the direct causes but also by other crossed effects. Therefore, it was necessary to construct two additional matrixes. Table 4 shows the relationship between the pillars. It is obvious that each sector is totally (1) related to itself; however, the incidence on other sectors is not symmetric.

Table 4. Relationship matrix between innovation pillars

	IN	HCR	INF	MS	BS
IN	1	0,4	0,6	0,5	0,7
HCR	0,7	1	0,8	0,4	0,9
INF	0,4	0,6	1	0,3	0,5
MS	0,3	0,3	0,3	1	0,8
BS	0,5	0,6	0,6	0,7	1

On the other hand, we need to assess the relationship that the *effects* (innovation capabilities of business) have on each other. Table 5 summarizes the information gathered.

Table 5. Relationship matrix between innovation capabilities

	IS	KM	PrM	PoM	ID	OE	ED
IS	1	0,4	0,4	0,6	0,4	0,4	0,8
KM	0,5	1	0,4	0,4	0,5	0,6	0,8
PrM	0,5	0,4	1	0,6	0,7	0,6	0,4
PoM	0,4	0,5	0,4	1	0,6	0,5	0,7
ID	0,7	0,6	0,6	0,7	1	0,6	0,8
OE	0,6	0,6	0,4	0,6	0,7	1	0,8
ED	0,5	0,7	0,5	0,6	0,9	0,6	1

We now proceed to generate the max-min (°) composition between the direct cause–effect matrix and the cause–cause matrix. Table 6 shows the first order incidences.

Table 6. First order incidence matrix

	IS	KM	PrM	PoM	ID	OE	ED
IN	0,7	0,6	0,5	0,7	0,5	0,6	0,7
HCR	0,7	0,6	0,6	0,7	0,8	0,6	0,8
INF	0,6	0,6	0,6	0,6	0,6	0,5	0,6
MS	0,7	0,3	0,5	0,7	0,3	0,5	0,7
BS	0,7	0,6	0,6	0,7	0,6	0,5	0,7

Finally, we proceed to perform the max-min (°) composition with the relationship matrix between innovation capabilities. We then obtain the final matrix, i.e., the second order incidence matrix. Table 7 shows the second order incidences.

Table 7. Second order incidence matrix

	IS	KM	PrM	PoM	ID	OE	ED
IN	0,7	0,7	0,5	0,7	0,7	0,6	0,7
HCR	0,7	0,7	0,6	0,7	0,8	0,6	0,8
INF	0,6	0,6	0,6	0,6	0,6	0,6	0,6
MS	0,7	0,7	0,5	0,7	0,7	0,6	0,7
BS	0,7	0,7	0,6	0,7	0,7	0,6	0,7

5 Discussion

The forgotten effects theory allows for the analysis of the direct and indirect incidence that each of the innovation pillars have on the innovation capabilities. Table 8 shows the absolute difference between the direct effect and the second-order incidence matrix.

Table 8. Total cumulative forgotten effect

	IS	KM	PrM	PoM	ID	OE	ED
IN	0,2	0,5	0,1	0,2	0,6	0	0
HCR	0	0,1	0	0	0	0,1	0
INF	0,1	0,1	0,2	0,3	0,2	0,3	0,1
MS	0,1	0,1	0	0,1	0,2	0,2	0
BS	0	0,1	0,1	0	0,3	0,1	0

It is evident that the influence of the variables: institutions, infrastructure and sophistication of markets due to the total indirect effect produced over the innovation capabilities. Table 9 gathers the cumulative forgotten effect obtained by the max-min compositions performed.

Table 9. Forgotten effect incidence

Cause	Effect	Initial value	Cumulative value	Forgotten effect
Institutions	Innovation strategy	$\mu_i(IN \rightarrow IS) = 0.5$	$\mu_a(IN \rightarrow IS) = 0.7$	0.2
Institutions	Knowledge management	$\mu_i(IN \rightarrow KM) = 0.2$	$\mu_a(IN \rightarrow KM) = 0.7$	0.5
Institutions	Portfolio management	$\mu_i(IN \rightarrow PoM) = 0.5$	$\mu_a(IN \rightarrow PoM) = 0.7$	0.2
Institutions	Internal drivers	$\mu_i(IN \rightarrow ID) = 0.1$	$\mu_a(IN \rightarrow ID) = 0.7$	0.6
Infrastructure	Project management	$\mu_i(INF \rightarrow PrM) = 0.4$	$\mu_a(INF \rightarrow PrM) = 0.6$	0.2
Infrastructure	Portfolio management	$\mu_i(INF \rightarrow PoM) = 0.3$	$\mu_a(INF \rightarrow PoM) = 0.6$	0.3
Infrastructure	Internal drivers	$\mu_i(INF \rightarrow ID) = 0.4$	$\mu_a(INF \rightarrow ID) = 0.6$	0.2
Infrastructure	Organization and structure	$\mu_i(INF \rightarrow OE) = 0.3$	$\mu_a(INF \rightarrow OE) = 0.6$	0.3
Market sophistication	Internal drivers	$\mu_i(SM \rightarrow ID) = 0.5$	$\mu_a(IN \rightarrow ID) = 0.5$	0.2
Market sophistication	Organization and structure	$\mu_i(SM \rightarrow OE) = 0.4$	$\mu_a(IN \rightarrow OE) = 0.6$	0.2
Business sophistication	Internal drivers	$\mu_i(BS \rightarrow ID) = 0.4$	$\mu_a(BS \rightarrow ID) = 0.7$	0.3

By performing a deep analysis of the variables' institutions, it is observed that experts have decided to assign a 0.1 direct effect over internal drivers; however, the interaction of the variable external drivers generates a total cumulative effect of 0.7. Therefore, a 0.6 total effect is first omitted and, thus, forgotten. This element needs to be reconsidered in order to properly assess the effect that it might bring to the model.

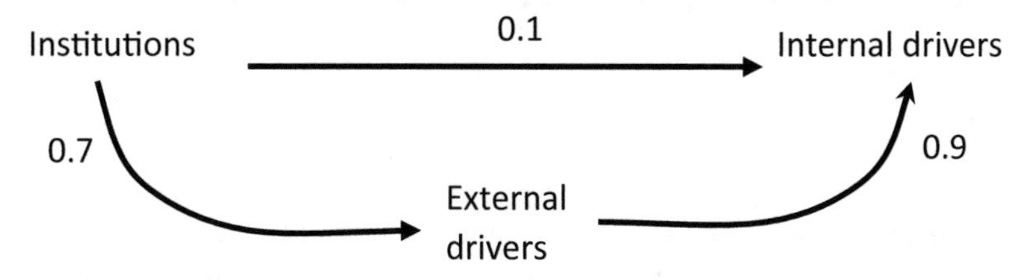

Institutions also exhibit a low direct incidence (0.2) over the variable knowledge management. However, the indirect cumulative effect that external drivers exert over the initial value makes the forgotten effect 0.7 (i.e., a total indirect effect of 0.5).

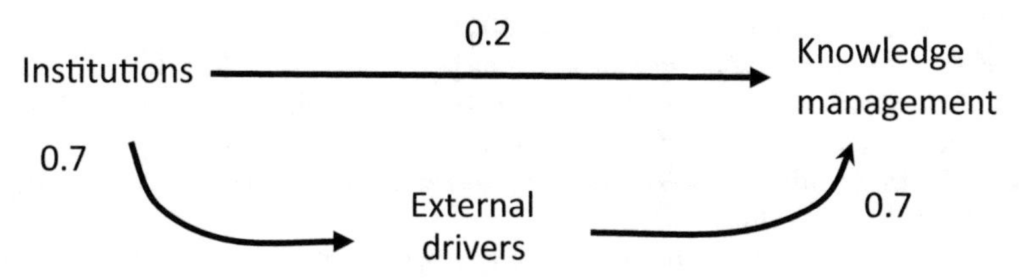

6 Conclusions

The innovation index in Mexico, developed by the Venture Institute, the National Council for Science and Technology and the Ministry of Economy, contributes to the generation of programs and policies that fosters innovation.

Through the theory of forgotten effects, taking into account the main innovation pillars considered by the variables model and innovation capabilities, it has been shown that some effects have not been fully considered in the first cause-effect matrix.

Some variables, such as institutions and infrastructure, have not been fully addressed since the cumulative effect of other variables, such as internal drivers, knowledge management and organization and structure, exerts positive effects that need to be considered to better direct the innovation efforts of the model.

This study is a first analysis on how a better structure of the programs supporting the innovation system of a country can be optimized by analyzing the indirect cumulative effect that the variables exert on each other.

There are some limitations to this study. To begin with, it is necessary to consider a wider pool of experts to assess the valuation matrixes. Additionally, better tools for interpreting the experts' opinions are required. Finally, further research is needed to assess the limitations of applying the theory of expertons and to gather more information, which can lead to more accurate results.

Acknowledgements. This paper is part of the Project supported by "Red Iberoamericana para la Competitividad, Innovación y Desarrollo" (REDCID) project number 616RT0515 in "Programa Iberoamericano de Ciencia y Tecnología para el Desarrollo" (CYTED).

References

Alfaro-García, V.G., Gil-Lafuente, A.M., Alfaro Calderón, G.G.: A fuzzy methodology for innovation management measurement. Kybernetes **46**(1), 50–66 (2017)

Adams, R., Bessant, J., Phelps, R.: Innovation management measurement: a review. Int. J. Manage. Rev. **8**(1), 21–47 (2006)

Comité Intersectorial para la Innovación: Programa Nacional de Innovación (2011). http://www.economia.gob.mx/files/comunidad_negocios/innovacion/Programa_Nacional_de_Innovacion.pdf. Accessed 10 June 2018

Cornell University, INSEAD, WIPO: The Global Innovation Index 2017: Innovation Feeding the World, Ithaca, Fontainebleau, and Geneva (2017)

Gil-Lafuente, A.M.: Fuzzy Logic in Financial Analysis. Springer, Berlin (2005)

Geoffrey, J.: Innovation, Growth and Competitiveness Dynamic Regions in the Knowledge-Based World Econo. Springer, Heidelberg (2009)

Kaufmann, A.: Introduction à la théorie des sous-ensembles flous à l'usage des ingénieurs (fuzzy sets theory). Masson (1977)

Kaufmann, A., Gil-Aluja, J.: Modelos para la investigación de efectos olvidados. Milladoiro, Vigo (1988)

OECD: "Innovación para el desarrollo de América Latina", Perspectivas económicas de América Latina 2015: Educación, competencias e innovación para el desarrollo. OECD Publishing, Paris (2014)

Venture Institute: Índice Nacional De Innovación (2013). http://www.fec-chiapas.com.mx/sistema/biblioteca_digital/indice-nacional-innovacion-2013.pdf. Accessed 15 June 2018

Digital Assets Horizon in Smart Cities: Urban Congestion Management by IoT, Blockchain/DLT and Human Reinforcement

Ferran Herraiz Faixó(✉), Francisco-Javier Arroyo-Cañada, María Pilar López-Jurado, and Ana M. Lauroba Pérez

Department of Business Administration, University of Barcelona, Av. Diagonal 690, 08034 Barcelona, Spain
ferranherraiz@ub.edu

Abstract. Cities are the future of humanity and for the coming majority, their next great home. This rapid urbanization of the planet presents great challenges and, at the same time, great opportunities. On the one hand, the concentration of people in urban spaces provides numerous economies of scale but, on the other hand, also implies the potential generation of negative externalities that must be managed by a "smarter city" to find a balance. Mobility in cities, the fact that moving in a vehicle (fuel, electric, shared, autonomous), public transport or other means is becoming a common, difficult and complex congestion problem for cities, has negative economic consequences. This enormous pressure for efficiency and the search for solutions are making significant inroads into the incorporation of ICT in smart cities and subsequent large-scale digitization to resolve this issue and other setbacks. This deployment needs to be conducted with a broad perspective incorporating economics and human concepts such as digital assets or citizen incentives in city strategies because in cities, people and resources are key players.

Using a forgotten effects methodology, the present exploratory work shows a new model of city congestion management deploying the possibilities offered by some disruptive technologies such as the Internet of Things (IoT), blockchain/distributed ledger technology (DLT) and token economy, combined with a human capital aspect such as a reinforcement theory. The potential possibilities that these concepts have in future mobility and other fields are large in terms of moving towards a more sustainable, decentralized and intelligent management enabling an authentic digital jump.

Keywords: Internet of Things · Distributed ledger technology · Blockchain · Smart cities · Congestion pricing · Motivation

---

This research has been possible thanks to the specific agreement between the Department of Business of the University of Barcelona and Telefónica Móviles España SA "Study and analysis of opportunities for improvement of services to citizens and businesses so that smart cities and/or other public organizations can achieve with the use of Big Data" within the framework agreement of collaboration for the development of joint initiatives formalized between the University of Barcelona and Telefónica S. A.

J. C. Ferrer-Comalat et al. (Eds.): MS-18 2018, AISC 894, pp. 63–82, 2020.
https://doi.org/10.1007/978-3-030-15413-4_6

1 Introduction

Currently, 55% of the world's population lives in urban areas. It is expected that in 2045 there will be 6 billion people living in cities, which will increase the current figure 1.5 times. Far from stagnating and looking beyond, it is expected that by 2050 the number of urbanites will reach 68% of the total world population (The World Bank 2018). This rapid urbanization that we are facing presents great challenges and great opportunities. The concentration of people in urban spaces provides numerous economies of scale, but at the same time, it implies negative externalities that must be managed including city congestion. This enormous pressure for urban space efficiency is opening the way to the incorporation of information and communication technology (ICT) in smart cities (SC), the Internet of Things (IoT) is especially relevant as a source of cities' efficiency management, and a balance needs to be considered between the technological aspects and human capital (Perera et al. 2014; Gil-García et al. 2016). It is not only about maximizing city digitization. It should be a step beyond the pure connectivity of the smart city 1.0 models and move towards optimization in decision making formulas that offer a quality of life for citizens and visitors based on a greater collaborative and consultative effort constructed on three objectives: to achieve a connected city, an intelligent economy and an innovative government where all the actors work together. Despite the fact that technology will play an important role in the future, we must again think that there are other "soft" factors such as citizen participation, collaboration and the reward of the intangible to be take into account in cities because one of the fundamental objectives of technology is to make better use of public resources, increase the quality of services offered to its citizens and, at the same time, improve the quality of life by stimulating the innovation process in the urban ecosystem (Cooray et al. 2017; Zanella et al. 2014, Giudice et al. 2011). Today's cities have become different systems characterized by abundant and continuous interconnections of their citizens, business, transportation, networks and communications.

The adjective "smart" added to the word cities can be traced back to the idea of the "smart growth movement" of the late 1990s whose promoters advocated new policies for urban planning. Subsequently, this concept was adopted by a large number of technology companies for the application of ICT complex systems (Harrison et al. 2011). Current reality leads us to the fact that there is no common definition regarding the "smart City" concept; however, there would be some consensus that these are characterized by a sharp use of ICT, although the deployment of these technologies does not necessarily guarantee a better city.

From a technological outlook, the literature review notes two major ICT roles in cities: on the one hand, there would be a technological role to improve productivity through the automation of routines and processes favouring better decision making. On the other hand, a body of studies would appear to place the focus on the human capital role, improving liveability by supporting and motivating the development of talent, participation and education. Thereby, there are two different views, one focused on optimization with the best urban intelligence management resources or "top-down" and the other focused on a "bottom-up" position where the city provides access to data and citizens make their own decisions. In this way and depending on the application and importance of ICT, two domains emerge: "hard" (with a greater preponderance of the application of new technologies in the fields of energy, resource management, environment, transport and mobility, residential, health, and safety) and "soft", where the ICT role is more limited- education, social inclusion, governance, economy, and innovation (Neirotti et al. 2013). Thus, there is no doubt that the deployment of ICT in cities is having a large impact on how a city is planned and organized and, more importantly, how the networks are established between the subjects and objects, from the social aspects to the economic ones - mobility, connectivity, consumers, commerce, taxes, resources, management, and collaboration (Batty et al. 2012).

Going beyond that, for the most advanced cities, the smart city 1.0 phase (connectivity ecosystems) is evolving towards the smart city 2.0 phase, seeking to move beyond infrastructure interconnection towards an optimization in the decision-making process through all actors' data management (governments, businesses and residents), offering a higher quality of life for citizens and visitors, greater economic competitiveness and an increase in sustainability and environmental awareness (Ottawa 2017; Eggers et al. 2018). This evolution must be based on new forms of collective intelligence (Malone et al. 2009) where governments push the creation of inclusive and open environments and platforms. This evolution requires a greater intelligence function to improve the coordination of different elements ensuring private, safety and transparent bidirectionality of information between administrations and citizens. Debate between what type of SC governance should have (centralized or decentralized) that allows efficiency, security, privacy and transparency is actually open.

The objectives of this paper are first rethink one of the most SC pressing problems, that is, mobility and, in particular, traffic congestion management and their externalities exhibiting smart congestion management (SCM) model; second, open a new citizen's role in cities administration by considering their decisions making process with more participation through truly end-to-end digitization system that combine ICT and human capital. It is argued that any proposed solution in the SC congestion issue should be carried out under a holistic approach that integrates the idea of the different typologies of smart city functions - smart people, smart economy, smart mobility, smart environment, smart living and smart governance (Batty et al. 2012). Therefore, delving into formulas that increase such integration, it is postulated that ICT advantages, in particular offered by IoT and blockchain/DLT (distributed ledger technology) and token economy, in addition with the promotion of the citizen reinforcement, has a superior solution in order contribute to solve the urban congestion issue in cities.

This paper is organized as follows: in Sect. 2, the theoretical background is described. In Sect. 3, the methodology is presented, followed in Sect. 4 by the results. Finally, Sect. 5 summarizes the most important conclusions of this paper.

2 Theoretical Background

2.1 Smart Cities Mobility

Cities move. Citizens and goods travel around cities by some type of vehicles from end to end in a frenetic nonstop motion through the combination of driving private or commercial vehicles, shared vehicles and public transport. Behind this increasing mobility, four new trends are observed that will mark the future, changing the way that we understand mobility -electrification, autonomy, connectivity and sharing- (Hannon et al. 2016).

Although the mobility concept is evolving, the underlying architecture needs to change. Consider some negative inefficiencies: in Boston, more than 40% of the vehicles during rush hour have only one occupant or vehicles are being used whose weight is 20 times that of the occupant; there is a misunderstanding with electric vehicles and the concept "zero emissions", which is not the same as "zero carbon emissions"; or in cities such as London, 24% of the urban area is allocated to roads, and this percentage reaches 40% in some cities in the US (Sumantran et al. 2018). City mobility has substantial problems in terms of difficult management such as management associated with pollution, climate change, urban congestion or energy consumption as cities consume up to 2/3 of the world's energy and account for more than 70% of greenhouse gas emissions (The World Bank 2018).

Specifically, in terms of mobility aspects, understanding how the elements combine and interact in a city is truly complex and depends on the situation of each city. In this way, there are 10 essential factors grouped into three blocks to understand city mobility in a clear and efficient way (Bouton et al. 2015): 1- depending on medium typology (private vehicle, walking or cycling, public transport, or new transport models); 2- according to the underlying system form (policies and regulations, urban design and land use, preferences and behaviours of consumers); and 3- the type of facilitators (financial, technological, or business models). Combinations of these factors will determine the basic mobility characteristics of each city as a unique case due to the concrete combination of factors.

The Horizon 2020 Program of the European Union includes objectives related to urban mobility and preferences within the Specific Program of SC4 "Intelligent, ecological and integrated transport" (European Commission 2017). This section is structured into four main types of activities: (a) efficient transportation in terms of resources that respect the environment; (b) better mobility, less congestion, and more security; (c) global leadership for the European transport industry; and (d) socio-economic and behavioural research and prospective activities for the formulation of policies to face the challenges posed by transport.

It is clear that sustainable mobility is a very important challenge, but a more ecological technology is not the solution. Having a higher vehicle fuel efficiency, the incorporation of electric vehicles (EV), an increase in the sharing economy or the appearance of autonomous vehicles (AV) adapted to consumer demand can become a potential for an increase in vehicle traffic rather than a decrease (Hahn 2011). Consider, for example, the evolution of commercial vehicles (CV) that handle home deliveries generated from commerce online and whose demand in cities is increasing significantly. It is expected that by 2025, approximately a quarter of American cities' population will use rapid delivery services on the same day (Bouton et al. 2017).

One of the most devastating city mobility effects is the congestion that results from it, considering this effect an aspect of inefficiency. It is estimated that the congestion impact on GDP (gross domestic product) in most cities is 2 to 4% in terms of waste of time or energy and the cost increase of doing business (Bouton et al. 2015). Thus, it is estimated that in the United States alone, congestion costs in cities will increase by 20% from 160 b $ in 2014 to 192 b $ in 2020 (Bouton et al. 2017).

The problems behind urban congestion are multiple. Focusing on productivity, the amount of time wasted in traffic is alarming and increasing. For example, Mexico City led the congestion index at 66% for 2016 (Tomtom 2016) - extra time needed for a vehicle to make the same journey on a congested road compared to a fluid traffic situation. Regarding this perspective, a worldwide snapshot by countries in 2017 provided by an INRIX consultant (INRIX 2017) say that Thailand, Indonesia, Colombia, Venezuela, Russia, the U.S., Brazil, South Africa, Turkey, and the U.K., in this order, are the "top 10" congestion countries ranked. The effects of this problem are dramatic because the costs incurred by these countries have a great impact on productivity in terms of GDP. Thus, as the Cookson CEO of INRIX (INRIX 2017) advises, "*Congestion costs the U.S. hundreds of billions of dollars, and threatens future economic growth and lowers our quality of life...*"

With this perspective, there are some solutions that are being put in place to face this great challenge. A review of the literature offers several proposals. First, we find citizen journey customization on a connected transport network by multimodal solutions (users can choose the best travel option at any time through the incorporation of ICTs and intelligent driving). Second, there has been a focus on accessibility and increasing the proximity concept (bringing the things we need more where we need them), eliminating the need to move through better urban planning and local production or by increasing technology penetration - teleservices (Zielinski 2012). Third, some cities are deploying driving restrictions or limitation in certain types of vehicles - Hamburg prohibits the circulation of the oldest and most polluting diesel vehicles in certain areas and streets (Europapress 2018). Fourth, restrictive measures are being implemented (prohibiting the driving of cars with odd or even number on alternate days is adopting in Madrid, Mexico, Beijing, Bogota, Caracas, and Santiago de Chile), and finally, there are toll initiatives (congestion pricing) where the vehicles that want to access city centres during in rush hour must pay for the privilege - Singapore, London, Stockholm, Milan, and Gothenburg. The final measure is supported by a recent study about a congestion pricing viability tax to avoid traffic jams and pollution in large cities (Fageda et al. 2018).

It is time to explore new opportunities if cities want to change mobility habits. Mobility outside the car should be more attractive for citizens. This requires attractive public transport accessible to all users (elderly, disabled, and parents with young children) where the mobility experience is positive with the existence of attractive transport nodes and enabling green or blue corridors to increase the pleasure of going on foot or by bicycle (Hahn 2011).

Cities and their leaders are introducing some measures to integrate mobility alternatives, for example, multimodality promotion with harmonized tariffs systems, new payment systems and the integration of flexible schedules, all unified through Internet applications close to the users.

These particular measures, in some way, are contributing greatly to better mobility and a reduction in urban congestion, but considering citizen habits as a cornerstone, all of these actions are not enough. In this way, looking for a better resources combination, mobility and congestion solution could be accelerated through a better arrangement of digital technology, in particular those offered by the IoT, blockchain/DLT and token economy, combined with citizen reinforcement theory, encouraging habit changes.

2.2 Digitization by Internet of Things, Blockchain/DLT

Enormous pressure towards SC efficient management is accelerating the use of ICT, discovering proper solutions. Among many other technologies behind the ICT concept, currently there are two major phenomena, the IoT (Internet of Things) and the blockchain/DLT (distributed technology ledger), revolutionizing the way we understand cities.

IoT earns more attention for its applicability in smart cities from different perspectives including social, economic, and technological ones. Cities can manage their resources more efficiently from these points of view, offering new value solutions that make the decision processes swifter (Perera 2014). IoT allows people and things to connect at anytime, anywhere, with anything and anyone that ideally uses any route/network and any service (Vermesan 2009). For example, the Telefónica Company (TelefonicaIoT 2018) notes, "*The hours lost in the long traffic jams in cities will be able to be invested in what truly matters thanks to the Internet of Things.*"

Blockchain/DLT allows for the appearance of the so-called "crypto-cities", transforming how cities coordinate and organize. A crypto-city is one that strives to open and decentralize city data using crypto-economy tools and technology (Potts et al. 2017). Thus, although the Blockchain technology is a remarkably transparent and decentralized way of registering lists and transactions, this will not be the solution to all city problems. It will have a large impact in many areas. In particular, its deployment in the sphere of the public sector will affect a large number of activities. According to the European Parliament (Boucher et al. 2017), "*... Blockchain technologies can provide new tools to reduce fraud, avoid errors, cut operating costs, boost productivity, improve compliance and improve accounting.*"

From another angle, IoT and the Blockchain/DLT deployment can contribute to enlarging the programmable economy or intelligent economic systems that allow cities to manage the production and consumption of goods and services in a different way. These technologies would form new and diverse scenarios of value exchange, both monetary and non-monetary (Gartner 2015). Overwhelmingly, this disruption allows for taking a further step towards IoT monetization, and therefore, it will be possible to convert urban physical or intangible assets (such as urban space or traffic congestion) as a digital asset, opening the possibility to creating new business models of management and revolutionaries' citizens and city interactions in a transparent and decentralized way.

Thus, these technologies will help to improve how cities and citizens connect (IoT) and how they are coordinated and organized (blockchain/DLT). However, if we want to change mobility habits, we would need to add the human behaviour factor. Tomorrow's cities (SC 2.0) are looking for positive citizens' and visitors' experiences, and they need to consider how to motivate them. The theory of motivation investigates why people vary their level of motivation, as well as determine the reasons why people are motivated.

2.3 The Human Factor. Incentives and Motivation

City congestion cannot be solved by Adam Smith's invisible hand because it fails to find an efficient result for those who are not part of the transaction but suffer the costs and benefits of the externalities, that is, the group of citizens; hence, the appearance of the so-called Pigouvian rates (Mankiw 2009). Therefore, we could make use of some factors beyond the technological and economic ones such as human reinforcement and motivation in order to stimulate and involve citizens in congestion resolution.

The theory distinguishes among diverse types of motivation in terms of the different underlying attitudes, reasons and objectives that give rise to the action. The clearest distinction is between intrinsic motivation - doing something because it is intrinsically interesting or pleasurable - and extrinsic motivation - doing something because one is externally forced to do it or because the action will lead to some type of reward or result (Ryan et al. 2000).

One of the ways to provide motivation is through the application of reinforcement theory, also known as learning theory (Skinner 1953). This refers to the stimuli used to produce desired behaviours. B. F. Skinner argued that for motivation, it is not necessary to understand individual needs, as proposed by the theories of the content of motivation. Skinner proposed that the administrator needs to only understand the relationship between behaviours and their consequences to create conditions that encourage desirable behaviours and discourage undesirable ones. Behaviour is learned through its positive or negative consequences. Reinforcement theory is widely used in organizations as an instrument to increase or decrease employees' behaviour and as a key factor to motivate staff (Wei et al. 2014). This appreciation could be used in city management in order to motivate citizens and reinforce or dissuade their mobility habits.

Combining these digital and disruptive technologies (IoT, blockchain, and DLT) as facilitators of citizens incentive factors through increasing their participation in city decision making as motivators transforms the city vision from "car-centric", anchored in the years 1920–30. to "people-centric", a vision based more on the idea of walking, riding, cycling or using public transport, which is greatly linked to cities' appropriate future concepts (WEF 2015). Therefore, positive and negative citizen reinforcements are fundamental points in the treatment of the congestion problem, and it becomes central but not the only way to address it in a different way.

Based on the above, the following hypotheses are presented for the SCM smart congestion management model:

H1- Negative incentives (congestion pricing, CP) applied to (private and commercial) unfriendly vehicles contributing to better mobility in cities.

H2- Positive incentives (uncongestion pricing, UP) applied in friendly vehicles contributing to better mobility in cities.

Friendly or unfriendly vehicles have been defined regarding the urban transport environmental impact in developed countries according to their efficiency in terms of respecting the environment while minimizing the impact on climate greenhouse gases (CHG), (WHO 2011): 1-Unfriendly: 61–170 max Co_2-eq emissions/passenger by travel (-gasoline, diesel, electric- cars, motorbike, vans, trucks, coach). 2-Friendly: 25–60 max Co_2-eq emissions/passenger by travel (bicycles, scooters, public transport-train, metro, bus).

2.4 Efficiency, Coordination and Transparency Functions

However, could all of these incentives be managed by current institutional systems ensuring, at the same time, efficiency and transparency process management? The backbone that will help with this characteristic will be the combination of IoT, blockchain/DLT technologies plus smart contracts and token economy elements because these digitals concepts will be crucial in city governance, as will be seen below.

Blockchain technology is found in the paper "Bitcoin: A Peer-to-Peer Electronic Cash System" by Satoshi Nakamoto (Nakamoto 2008) and refers to a particular way of organizing and storing information and transactions between peers (peer to peer). Blockchain is one type of a distributed ledger. Distributed ledgers use independent computers (referred to as nodes) to record, share and synchronize transactions in their respective electronic ledgers instead of keeping data centralized as in a traditional ledger (The World Bank 2018). We can define this blockchain technology as *"a list of irrevocable transaction records, cryptographically signed, shared by all participants in a network, in a peer-to-peer manner"* (Gartner 2018).

This technology enables the second generation of the Internet, that is, through it, we go from "Internet of information" to "Internet of value" radically changing the operational of the economy and of institutions (Tapscott 2017). Analysing this technology scope within the digitalization sphere using Gartner's DD (Digital Disruption) scale, currently this technology is in the DD1 stage or initial phase of disruption, and it will reach DD3 stage or transformation stage in 2023 estimating that the value generated by this in the economy world will be 176 b $ in 2025 (Kandaswamy et al. 2018).

Blockchain/DLT technology drastically reduces the cost of trust and allows for imagining new business models and opens the possibility to deploying new forms of collaboration at the same time. As a source of trust, it can expand organizations and institutions digital transformation degree where they can share processes among all actors, which is possible because this technology makes the information contained within immutable, safe, shared and consensual (HBR 2017). Additionally, this registry allows for expanding the smart contracts concept or intelligent contracts, a concept attributed to Nick Szabo (Szabo 1996). He observed that contracts between parties are one of economic capitalism's pillars, and they are similar to the computational programs, as they are representations in the form of code of the same processes while allowing to certify them. In other words, organizations and institutions can agree in such contracts as to how they want their business or public services to be carried out in a consistent and automated manner, keeping participant interactions' records in a transparent manner at all times. This technology allows for a better decision-making process either between people or between machines without human intervention.

It is usually assumed that a smart city tends to increase its bureaucracy. One of the most important characteristics of this technology is that it allows for reducing, in a very consistent manner, the so-called "transaction costs" defined in Coase's work "The Nature of the Firm" (Coase 1937), where the choice between the organization or the market is essentially an exercise in minimizing costs. These costs would include the cost of organization, negotiation, monitoring, information and coordination (Potts 2017), digital technologies clearly can reduce many of these costs and blockchain/DLT technology tends to reduce decisions and many coordination functions could be replace by the smart contracts software.

Therefore, an approach of city management and more specifically citizen's mobility and urban congestion treatment and control by transaction cost theory, can open new institutional possibilities in order to reduce transaction costs, control, monitoring expenses and give them the possibility to achieve more efficient data coordination.

Thus, blockchain/DLT technology allows for a strong move towards connected mobility by using smart contracts and digital wallets in all types of vehicles. This innovation means carrying out all transfers through the Blockchain/DLT, which recognizes its activation under certain conditions (Sandner et al. 2017) that institutions can define. In this way, unfriendly vehicles can make certain token transfers automatically to pay congestion charges (congestion pricing) for accessing a specific city area.

In contrast, friendly vehicles (and their owners) can obtain positive incentives (Uncongestion Pricing) if they access the same area in a more sustainable manner, in the form of this token and using them such that with public transport discounts or paying local tax or commerce coupons, they can exchange tokens by fiat money, all with minimal human interaction.

Thus, IoT allows people and things to be connected anytime, anywhere, with anything and anyone, ideally using any route/network (Vermesan 2009). This technology can be used in order to know which vehicles or individuals access certain city areas with high urban congestion while maintaining their privacy. In addition, token economy and blockchain activate the possibility to deploy the digital asset concept (digital asset market in representation of the physical assets), preventing duplication through the generation of smart contracts or protocols that facilitate, verify and execute the terms of the city contract, all thanks to the fact that tokens or units of value can be employed to make the payment (Kandaswamy et al. 2018).

Thus, through these technologies, it is possible to create a new city congestion scenario to manage and negotiate access to any vehicle for specific areas, giving citizens positive or negative incentives and dissuading unfriendly vehicles from use while encouraging friendly ones such as public transport or bicycles.

Blockchain/DLT and IoT applied to the institutional administration become and excellence technologies in the governance and decentralization way. Their impact on productivity and efficiency can be extremely beneficial (Davidson et al. 2016). In particular, blockchain/DLT can be a good tool to reduce fraud, avoid errors, reduce costs, promote productivity and strengthen compliance in public services. Future applications can provide some benefits such as tax payment system, identification tool, benefits distribution platform, issuance of local digital currencies, or property registers (UK Government Chief Scientific Adviser 2016).

This type of technology has been applied in different automotive use cases such a vehicle identification, payment of tolls and EV charging (allowing automatic, transparent and frictionless payment), and it is offering some advantages such as decreasing complexity, no transaction fee and many implications of the machine-to-machine economy, modelling new business models by, for instance, IOTA utility token applications and the tangle possibilities in SC such as Taipei or Pangyo among others (Pauseback 2018).

Thus, considering the smart city dimensions under the European prism mapping smart cities in the EU (smart economy, smart people, smart mobility, smart environment, smart governance, and smart living), which considers that any improvement included in one or more of these dimensions is a sufficient reason to consider a city as intelligent (European Parliament 2014), the following hypothesis is presented:

H3- Negative incentives (congestion pricing, CP) applied in (private or commercial) unfriendly vehicles combined with positive incentives (uncongestion pricing) applied in friendly vehicles by token economy (congestion pricing digital asset) and anchored in the IoT and DLT technologies contribute to be better smart mobility (reduce city congestion, improve institutional efficiency and recover city productivity) and better SC, greater than H1 or H2.

With H1, H2 and H3, the smart congestion management SCM model is described in Fig. 1:

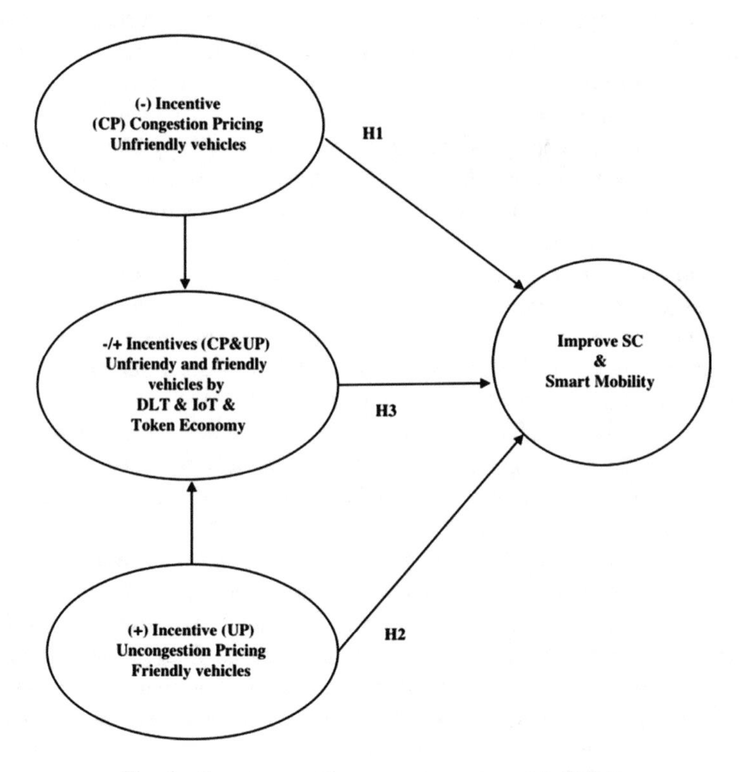

Fig. 1. Smart congestion management model (SCM)

3 Methodology

3.1 Incidents Analysis by Forgotten Effects

With the idea of determining the model's validity and calculating its overall effect, considering the direct and indirect incidents of each of the selected causes, the forgotten effects theory will be used (Kaufmann and Gil-Aluja 1988; Gil- Lafuente 2008).

First, the causes "c" and the effects "e" of the SCM model are defined in Table 1:

Table 1. Causes and effects in smart congestion management (SCM) model

Causes	Effects
c_1: (−) Incentives, Congestion pricing (CP), Unfriendly vehicles	e_1: Reduce traffic index (congestion level) e_2: Improve productivity (reduce waste of time) e_3: Improve environment (green areas and reduce noise/Co_2 pollution)
c_2: (+) Incentives, Uncongestion pricing, (UP) Friendly vehicles	e_4: Improve health (personal stress, related diseases, accidents) e_5: Increase public transport and friendly vehicle use (bicycle, electric skates)
c_3: (− and +) Incentives (CP&UP), Unfriendly and friendly vehicles by Blockchain/DLT & IoT & token economy	e_6: Improve local financial resources e_7: Increase institutional governance efficiency (reduce transaction cost and monitoring) e_8: Increase connectivity, network (digitization) e_9: Increase citizen participation (P2P, Apps) e_{10}: Built the backbone to deploy others digital assets (reward citizen behaviour)

To determine the incidences of the causes among themselves, the incidents between the causes and the effects, and the effect incidences between them, a survey was developed with experts to obtain an assessment with a scale of 0.0 to 1.0 (decimal scale) in order to know these incidents. In this way, the value of the incidence between the variables is assigned (1.0 value means significant and 0.0 value means not significant); it has been considered that the incidence of a variable on itself is total, so the assigned value is 1.0. In this way, three necessary matrixes must be calculated in order to obtain forgotten effects:

With C = ($c_{1.}$ c_2,…, c_n) being a set of causes and E = ($e_{1.}$ e_2,…, e_n) being a set of effects, the incidences of the elements of C on the elements of E, (c_i, e_j) = μ_{ij} can be valued in the interval [0, 1], with the assigned value being greater, the more intense the incidence is. Therefore:

$$\mu_{ij} \in [0,1]; c = 1,2,\ldots,n; e = 1,2,\ldots,m$$

Then, the following matrixes are built (Table 2):

$$\forall(\mathrm{c_i},\mathrm{c_k}) \in \left[\widetilde{\mathrm{C}}\right] : \propto_{\mathrm{ik}} \in [0,1]; \mathrm{i} = \mathrm{k} : \propto_{\mathrm{ii}} = 1$$

$$\forall(\mathrm{e_i},\mathrm{e_l}) \in \left[\widetilde{\mathrm{E}}\right] : \beta_{\mathrm{il}} \in [0,1]; \mathrm{j} = 1 : \beta_{\mathrm{jj}} = 1$$

Table 2. Matrix

1- Between causes matrix	$\left[\tilde{C}\right] = \begin{array}{c\|cccc} & c_1 & c_2 & \ldots & c_m \\ c_1 & \alpha_{11} & \alpha_{12} & \ldots & \alpha_{1m} \\ c_2 & \alpha_{21} & \alpha_{22} & \ldots & \alpha_{2m} \\ \ldots & \ldots & \ldots & \ldots & \ldots \\ c_n & \alpha_{n1} & \alpha_{n2} & \ldots & \alpha_{nm} \end{array}$
2- Between effects matrix	$\left[\tilde{E}\right] = \begin{array}{c\|cccc} & e_1 & e_2 & \ldots & e_m \\ e_1 & \beta_{11} & \beta_{12} & \ldots & \beta_{1m} \\ e_2 & \beta_{21} & \beta_{22} & \ldots & \beta_{2m} \\ \ldots & \ldots & \ldots & \ldots & \ldots \\ e_n & \beta_{n1} & \beta_{n2} & \ldots & \beta_{nm} \end{array}$
3- Between causes and effects matrix	$\left[\tilde{M}\right] = \begin{array}{c\|cccc} & e_1 & e_2 & \ldots & e_m \\ c_1 & \mu_{11} & \mu_{12} & \ldots & \mu_{1m} \\ c_2 & \mu_{21} & \mu_{22} & \ldots & \mu_{2m} \\ \ldots & \ldots & \ldots & \ldots & \ldots \\ c_n & \mu_{n1} & \mu_{n2} & \ldots & \mu_{nm} \end{array}$
4- Global matrix	$\left[\widetilde{M^*}\right] = \left[\tilde{C}\right]^{\circ}\left[\tilde{M}\right]^{\circ}\left[\tilde{E}\right] = \begin{array}{c\|cccc} & e_1 & e_2 & \ldots & e_m \\ c_1 & \omega_{11} & \omega_{12} & \ldots & \omega_{1m} \\ c_2 & \omega_{21} & \omega_{22} & \ldots & \omega_{2m} \\ \ldots & \ldots & \ldots & \ldots & \ldots \\ c_n & \omega_{n1} & \omega_{n2} & \ldots & \omega_{nm} \end{array}$

The matrix $\left[\widetilde{M}^*\right]$ will be the result of first making a convolution (°) between the matrix of causes and the matrix that includes the direct incidents of causes and effects:

$$\text{1st Convolution maxmin :}$$
$$\upsilon(c_n, e_m) = \vee(\alpha_{nn} \wedge \mu_{nm})$$

and, later, a second convolution between the resulting matrix $\left[\left[\tilde{C}\right]^{\circ}\left[\tilde{M}\right]\right]$ and the matrix that collects the incidents between the variables of the effects $\left[\tilde{E}\right]$.

$$2^{nd}\text{ Convolution maxmin :}$$
$$\omega(c_n, e_m) = \vee(\upsilon_{nn} \wedge \beta_{nm})$$

Finally, once the global incidents $\left[\widetilde{M}^*\right]$ have been obtained, if we subtract the direct incidents of the causes on the effects $\left[\tilde{M}\right]$, we can determine the indirect incidents of each one of the causes on the effects or forgotten effect matrix.

$$\left[\widetilde{M^*}\right] - \left[\widetilde{M}\right] = \begin{array}{cccccc} & e_1 & e_2 & \ldots & e_m \\ c_1 & \theta_{11} & \theta_{12} & \ldots & \theta_{1m} \\ c_2 & \theta_{21} & \theta_{22} & \ldots & \theta_{2m} \\ \ldots & \ldots & \ldots & \ldots & \ldots \\ c_n & \theta_{n1} & \theta_{n2} & \ldots & \theta_{nm} \end{array}$$

3.2 Sample

To be able to carry out the investigation, the project has been divided into two phases: in the first and current phase, we test the model with two SC experts, later introducing the acquired learning, and proceed to phase two, extending the experts number to 10 in the fields of institutional, technology, blockchain, urban planning, public transport, private sector, automotive industry, and commerce as well as academic.

The first phase model has been tested, carrying out two personal interviews with SC experts in the public administration and private enterprise sectors. The interview methodology started by explaining the model, and next, experts completed some matrixes (causes/causes ($\left[\widetilde{C}\right]$), causes/effects ($\left[\widetilde{M}\right]$) and effects/effects ($\left[\widetilde{E}\right]$). Then, the experts shared their impressions and noted the following:

1. The estimated time to complete the matrixes was approximately 1 h.
2. There were some conditioning issues in the matrix c/e and e/e answers (first five columns), so it was proposed to change their order.
3. Two different approaches emerged in the way of determining the matrix according to whether the expert belongs to the public administration or the private sector.
4. The public administration expert mentioned the importance of the slow tech adoption curve in the institutional environment.
5. The private sector expert mentioned that in public administrations some projects have difficulties depending on how many departments were involved with (with more department, there are more problems) and most important, large projects must require a long-term vision, members' commitments and strong political stability.
6. The experts agreed that the token economy and the digital assets phenomena can be exported to other SC projects beyond congestion management application.

4 Results

The variables used in this analysis are the same as those used in Table 1 of the previous section. With the intention of knowing the total incidences of the different types of causes on the variables (effects) of the smart congestion management SCM model, the forgotten effects methodology described in the previous section will be used.

$$\left[\widetilde{M^*}\right] = \left[\widetilde{C}\right]^{\circ}\left[\widetilde{M}\right]^{\circ}\left[\widetilde{E}\right]$$

The preliminary results obtained follow this structure:

First, the direct incidents between causes and the effects in SCM model are determined in Table 3. The values have been obtained as a result of applying an arithmetic average from the different scores of the experts.

Table 3. Incidents between causes and SCM variables $\left[\tilde{M}\right]$

$\left[\tilde{M}\right]$	e_1	e_2	e_3	e_4	e_5	e_6	e_7	e_8	e_9	e_{10}
c_1	0.20	0.20	0.20	0.20	0.40	0.45	0.55	0.10	0.10	0.10
c_2	0.60	0.60	0.60	0.60	0.85	0.20	0.30	0.25	0.40	0.80
c_3	0.75	0.75	0.65	0.60	0.85	0.50	0.80	0.75	0.65	0.80

Second, the incidences between the effects of the SCM model are collected in Table 4. The values have been obtained as a result of applying an arithmetic average from the different scores of the experts.

Table 4. Incidents between variables SCM variables $\left[\tilde{E}\right]$

$\left[\tilde{E}\right]$	e_1	e_2	e_3	e_4	e_5	e_6	e_7	e_8	e_9	e_{10}
e_1	1.00	1.00	1.00	1.00	0.60	0.10	0.10	0.40	0.50	0.35
e_2	0.30	1.00	1.00	1.00	0.70	0.80	0.40	0.40	0.35	0.25
e_3	0.20	0.45	1.00	1.00	0.25	0.40	0.20	0.30	0.35	0.30
e_4	0.10	0.65	0.40	1.00	0.80	0.55	0.50	0.40	0.45	0.30
e_5	0.90	0.70	0.90	0.45	1.00	0.35	0.35	0.25	0.40	0.40
e_6	0.45	0.40	0.50	0.40	0.70	1.00	0.85	0.50	0.50	0.40
e_7	0.35	0.70	0.50	0.20	0.45	0.80	1.00	0.90	0.60	0.50
e_8	0.70	0.75	0.50	0.35	0.60	0.70	0.85	1.00	0.85	0.85
e_9	0.50	0.85	0.65	0.50	0.80	0.50	0.70	0.65	1.00	0.85
e_{10}	0.85	0.80	0.65	0.35	0.65	0.75	0.75	0.65	0.85	1.00

Third, the incidences of the different types of causes among themselves are determined in Table 5. The values have been obtained as a result of applying an arithmetic average from the different scores of the experts.

Table 5. Incidents between SCM causes $\left[\tilde{C}\right]$

$\left[\tilde{C}\right]$	c_1	c_2	c_3
c_1	1.00	0.35	0.85
c_2	0.55	1.00	0.90
c_3	0.75	0.85	1.00

Fourth, we obtain the global incidence, both direct and indirect, through the "maxmin" first convolution process

$$\left[\widetilde{M^*}\right] = \left[\tilde{C}\right]^{\circ}\left[\tilde{M}\right]^{\circ}\left[\tilde{E}\right]$$

as shown in Table 6:

Table 6. Global incidence - total cause/effects on the SCM variables $\left[\widetilde{M^*}\right]$

$\left[\widetilde{M^*}\right]$	e_1	e_2	e_3	e_4	e_5	e_6	e_7	e_8	e_9	e_{10}
c_1	0.85	0.85	0.75	0.85	0.85	0.80	0.80	0.80	0.80	0.80
c_2	0.85	0.85	0.75	0.85	0.85	0.80	0.80	0.80	0.80	0.80
c_3	0.85	0.85	0.75	0.85	0.85	0.80	0.80	0.80	0.80	0.80

Once the global incidents have been obtained, if we subtract the direct incidences of the causes on the variables of the SCM model (which have been obtained in the first step), then we can know the indirect incidences of each of the causes on the variables of the SCM model by the second convolution process, as can be seen in Table 7:

Table 7. Forgotten effects in SCM model $\left[\widetilde{M^*}\right] - \left[\tilde{M}\right]$

$\left[\widetilde{M^*}\right] - \left[\tilde{M}\right]$	e_1	e_2	e_3	e_4	e_5	e_6	e_7	e_8	e_9	e_{10}
c_1	0.65	0.65	0.55	0.65	0.45	0.35	0.25	0.70	0.70	0.70
c_2	0.25	0.15	0.15	0.25	0.00	0.60	0.50	0.55	0.40	0.00
c_3	0.10	0.10	0.10	0.25	0.00	0.30	0.00	0.05	0.15	0.00

These results show that, directly or indirectly (global incidence), any cause raised in the smart city condition has a minimum level of 0.75 over 1.00.

It can be verified that cause c1 has the greatest forgotten effects with a camouflage of its global incidences on the model's effects, particularly for the effects e_8, e_9, e_{10} all with values 0.7 on 1.00.

It is observed that c_1: "(−) incentives, congestion pricing (CP) unfriendly vehicles", is the one that has a less direct impact on "reducing traffic index" indicators $e_1 = 0.20$ "productivity improvement" $e_2 = 0.20$ and "improving institutions efficiency" $e_7 = 0.55$. Progressively, it is observed that c_2: "(+) incentives, uncongestion pricing (UP) friendly vehicles" has more influence on these indicators. When analysing how the cause c_3 affects "(− and +) incentives (CP & UP) unfriendly and friendly vehicles by blockchain, DLT & IoT & token economy", it is verified that the direct affectation is still greater with a significant impact on "reducing traffic index" $e_1 = 0.75$, "productivity improvement" $e_2 = 0.75$ and "improving institution efficiency" $e_7 = 0.80$.

For example, if we analyse the effect that c_3 has on "reducing traffic index", if we consider only the direct effect (Table 3), then the incidence is 0.75, but if the cumulative effect is observed due to the existence of other interposed variables (Table 6), the effect increases up to the value 0.85. The isolated indirect effect is 0.10 on 1.00 or very low. The interposed effect that intervenes most importantly and that determines that the cumulative effect is maximum is the variable "public transport and friendly alternative" (Fig. 2).

At the same time, we can see how c_1 is the one with the greatest forgotten effects and following the same example, and its direct impact on the reduction of congestion level (Table 3) is 0.20. However, if the accumulated effect is observed due to the impact of the interposed variables (Table 6), then its effect increases to the value 0.85. The isolated forgotten effect is very relevant, 0.65 on 1.00. The effect interposed in this case "public transport and friendly vehicle use" explains a part and interposed cause c_3, and its interposed effects would explain the remainder (Fig. 2):

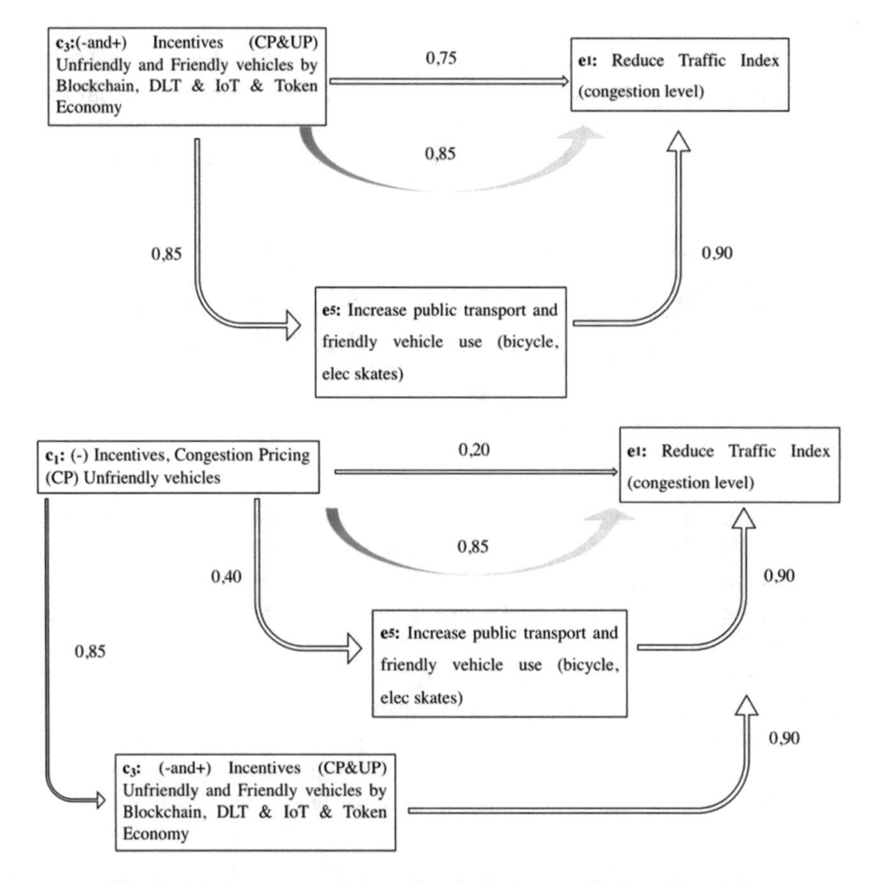

Fig. 2. Maximum cumulative effect in "reduce traffic index" variable.

Therefore, considering exclusively the direct effects of causes $c_{1.}$ c_2 and c_3 on the SCM model's variables, the effect on the improvement of congestion level and most of the variables that affect the level of welfare, efficiency and productivity, would be mainly determined from lower to higher incidence for causes $c_{1.}$ then cause c_2, and finally and with greater impact, cause c_3. In addition, if we consider the forgotten effects, c_1 is the one with the highest values, followed by c_2 and c_3 in this order, which indicates that the direct effect of causes c1 and c_2 is lower if they act in isolation as opposed to intercepting c_3, so provisionally we could confirm the hypotheses H1, H2 and H3 of the SCM model.

5 Conclusions

The results presented by this exploratory study would allow institutions and organizations to take more effective measures to improve SC mobility and, in particular, to improve urban congestion that is problematic from a broader point of view that includes technological factors and human capital elements. The fundamental fact of knowing the direct and indirect effects allows for having very valuable information that is often omitted in decision making. Although any of the three measures proposed in the SCM model influences the SC and citizen welfare, it is observed that the combined technological and human motivation effects (c_3), at the same time, have a provisionally greater impact on improving urban congestion, city productivity, institutional efficiency rather than congestion pricing (c_1) or "uncongestion" pricing (c_2) in isolation.

Most cities and their multiple actors are increasingly aware of the convenience of having an efficient and transparent city management model that encourages citizen participation, while deploying technology use although technology, by itself, does not solve the SC problems. Thus, solutions to major urban challenges require a particular vision, avoiding general technological recipes. In this sense, urban congestion problems (productivity, environment and human health) and solutions need to be approached by not only a digitization umbrella, pursued to win the digital indicators race but also mixing holistic and particular perspective to include others key variables such as human motivational aspect in order to balance externalities and shift urban models ahead from 1.0 to 2.0.

Presently, solutions to the urban congestion phenomena are partial, and as already mentioned, a greener technology, sharing economy or electric or autonomous vehicles incorporation would not solve the traffic density problem and could even have the opposite effect. Fortunately, the digital phenomena allow cities to evolve from the Internet of information to the Internet of value, opening up new possibilities that should be exploited by SC. This disruption will make it possible to exchange intangible things that citizens of a society consider valuable. Current and future projections show that people will be living in cities with growth, and as a consequence, tangible and intangible city resources will be scarce, including urban congestion and its negatives effect on productivity, wasted time, health and other issues, consequences that cities will need to manage. Nonetheless, all of these externalities can be digitized as digital externalities such as city digital asset/passive, offering the possibility to exchange them among citizens thanks to available technology (IoT, blockchain, DLT, smart contracts, and tokens). In contrast, adding positive and negative reinforcement incentives for citizens in the equation, will make end-to-end citizen value exchange different and more participatory, distributed, reliable, efficient and transparent formula in a different way such as the tax city institutional predatory perspective, avoiding a centralization mentality.

Thus, institutional roles and industry manager predispositions will play key role in the SC congestion solution. They will need to invest in a long-term vision because cross digital steps such as SCM are not easy to implement. There is a long path from a centralized, institutional city control to decentralized participation. These new approach flags will allow the incorporation of new rules, adding the possibility to create new intelligent and autonomous city management, configuring a new economy, new scenarios and even new social roles unthinkable until now.

As a limitation of this article, it should be noted that this study is in its first phase (work in progress). At the first moment, we have attempted to test the feasibility of the model to move on to the next phase. However, among other limitations, this paper does not discuss important technologies needed to implement an SCM model such as 5G, artificial intelligence, or big data.

As new areas of research, we propose to analyse the SCM model in a large SC are such as Barcelona, in addition to adding new causes and/or effects and even use other methodologies.

References

Batty, M., Axhausen, K.W., Giannotti, F., Pozdnoukhov, A., Bazzanio, A., Wachowicz, M., Ouzounis, G., Portugali, Y.: Smart cities of the future. Eur. Phys. J. Spec. Top. **214**, 481–518 (2012)

Boucher, P., Nascimento, S., Kritikos, M.: How blockchain technology could change our lives. Eur. Parliamentary Res. Serv., 24 (2017). https://doi.org/10.2861/926645

Bouton, S., Knupfer, S.M., Mihov, I., Swartz, S.: Urban mobility at a tipping point, pp. 1–23. McKinsey & Company, New York City (2015)

Bouton, S., Hannon, E., Haydamous, L., Heid, B., Knupfer, S., Naucler, T., Neuhaus, F., Nijssens, S., Ramanathan, S.: An integrated perspective on the future of mobility, Part 2: Transforming Urban Delivery (2017). Mckinsey.com. https://www.mckinsey.com/~/media/mckinsey/business%20functions/sustainability%20and%20resource%20productivity/our%20insights/urban%20commercial%20transport%20and%20the%20future%20of%20mobility/an-integrated-perspective-on-the-future-of-mobility.ashx

Coase, R.H.: The nature of the firm. Economica **4**, 386–405 (1937)

Cooray, M., Duus, R., Bundgaard, L.: These three factors are critical to the success of future cities. World Economic Forum (2017). http://amp.weforum.org/agenda/2017/09/these-three-factors-are-critical-to-the-success-of-future-cities

Davidson, S., De Filippi, P., Potts, J.: Disrupting governance: the new institutional economics of distributed ledger technology. SSRN (2016). https://doi.org/10.2139/ssrn.2811995

Eggers, W., Skowron, J.: Forces of change: Smart Cities (2018). Www2.deloitte.com. https://www2.deloitte.com/content/dam/insights/us/articles/4421_Forces-of-change-Smart-cities/DI_Forces-of-change-Smart-cities.pdf

Europapress: Hamburgo podría convertirse en la primera ciudad alemana que prohíbe circular a los vehículos diésel. europapress.es (2018). http://www.europapress.es/motor/coches-00640/noticia-hamburgo-podria-convertirse-primera-ciudad-alemana-prohibe-circular-vehiculos-diesel-20180521183834.html

European Commission: Horizon 2020: smart, green and integrated transport. Important Notice on the Second Horizon 2020 Work Programme, 129 (2017)

European Parlament: Mapping Smart Cities in the EU (2014). Europarl.europa.eu. http://www.europarl.europa.eu/RegData/etudes/etudes/join/2014/507480/IPOL-ITRE_ET(2014)507480_EN.pdf

Fageda, X., Flores-Fillol, R.: Atascos y contaminación en las grandes ciudades: Análisis y soluciones. Fedea Policy Papers (2018)

Gartner: Analysts Explore Next Phases of Digital Business at Gartner Symposium/ITxpo, 4–8 October 2015 in Orlando. https://www.gartner.com/newsroom/id/3146018

Gartner: Search for "Blockchain": Gartner IT Glossary. Gartner IT Glossary (2018). https://www.gartner.com/it-glossary/?s=Blockchain

Gil-Lafuente, A.M.: Incertesa i Bioenginerìa. Real Academia de doctors, Barcelona (2008)

Gil-Garcia, J.R., Pardo, T.A., Nam, T.: Smarter as the New Urban Agenda, pp. 1–19. Springer, Cham (2016). https://doi.org/10.1007/978-3-319-17620-8

Giudice, M., Straub, D.: IT and entrepreneurism: an on-again, off-again love affair or a marriage? MIS Q. **35**, iii–vii (2011)

Hahn, J.: Cities of tomorrow- challenges, visions, ways forward. European Union Regional Policy, 112 (2011). https://doi.org/10.2776/41803

Hannon, E., McKerracher, C., Orlandi, I., Ramkumar, S.: An integrated perspective on the future of mobility. McKinsey & Company, New York City(2016). https://www.mckinsey.com/business-functions/sustainability-and-resource-productivity/our-insights/an-integrated-perspective-on-the-future-of-mobility

Harrison, C., Donnelly, I. A.: A theory of smart cities. In: Proceedings of the 55th Annual Meeting of the ISSS – 2011. Hull, UK, pp. 1–15 (2011). https://doi.org/10.1017/cbo9781107415324.004

HBR: How Blockchain Will Accelerate Business Performance and Power the Smart Economy - Sponsor Content From Microsoft. Harvard Business Review (2017). https://hbr.org/sponsored/2017/10/how-Blockchain-will-accelerate-business-performance-and-power-the-smart-economy. Accessed 07 Nov 2018

INRIX: Los Angeles Tops INRIX Global Congestion Ranking| INRIX. INRIX - INRIX (2017). http://inrix.com/press-releases/scorecard-2017/

Kandaswamy, R., Furlonger, D.: Pay Attention to These 4 Types of Blockchain Business Initiatives (2018). Gartner.com. https://www.gartner.com/doc/3868969/pay-attention-types-Blockchain-business

Kandaswamy, R., Furlonger, D., Stevens, A.: Digital Disruption Profile: Blockchain's Radical Promise Spans Business and Society (2018). Gartner.com. https://www.gartner.com/doc/3855708/digital-disruption-profile-Blockchains-radical

Kaufmann, A., Gil-Aluja, J.: Models per la recerca d'efectes oblidats, Vigo. Milladoiro (1988)

Malone, T.W., Laubacher, R., Dellarocas, C.: Harnessing crowds: mapping the genome of collective intelligence. SSRN (2009). https://doi.org/10.2139/ssrn.1381502

Mankiw, N.G.: Smart taxes: An open invitation to join the pigou club. East. Econ. J. **35**, 14–23 (2009)

Nakamoto, S.: Bitcoin: A Peer-to-Peer Electronic Cash System (2008). Www.Bitcoin.Org9. https://doi.org/10.1007/s10838-008-9062-0

Neirotti, P., De Marco, A., Cagliano, A.C., Mangano, G., Scorrano, F.: Current trends in smart city initiatives: Some stylised facts. Cities **38**, 25–36 (2013)

Ottawa: Smart City 2.0. City of Ottawa. Planning, Infrastructure and Economic Development (2017). Documents.ottawa.ca. https://documents.ottawa.ca/sites/documents.ottawa.ca/files/smart_city_strategy_en.pdf

Pauseback, J. Modelling New Business Models with IOTA – IOTA (2018). https://blog.iota.org/modelling-new-business-models-with-iota-fadd53c6a192

Perera, C., Zaslavsky, A., Christen, P., Georgakopoulos, D.: Sensing as a service model for Smart Cities supported by Internet of Things. Trans. Emerg. Telecommun. Technol. **25**, 81–93 (2014)

Potts, J., Rennie, E., Goldenfein, J.: Blockchains and the Crypto City (2017). Degruyter.com. https://www.degruyter.com/view/j/itit.2017.59.issue-6/itit-2017-0006/itit-2017-0006.xml

Ryan, R.M., Deci, E.L.: Intrinsic and extrinsic motivations: classic definitions and new directions. Contemp. Educ. Psychol. **25**, 54–67 (2000)

Sandner, P., Wegner, L.: A Disruptive Technology for the Next Step of Digitization (2017). Digitalspirit.dbsystel.de. https://digitalspirit.dbsystel.de/en/a-disruptive-technology-for-the-next-step-of-digitization/

Skinner, B.: Science and Human Behavior, pp. 59–90. Macmillan, New York (1953). https://doi.org/10.1901/jeab.2003.80-345

Sumantran, V., Fine, C., Gonsalvez, D.: Faster, Smarter, Greener. The MIT Press, Cambridge (2018). https://mitpress.mit.edu/books/faster-smarter-greener

Szabo, N.: Smart Contracts: Building Blocks for Digital Markets (1996). Fon.hum.uva.nl. http://www.fon.hum.uva.nl/rob/Courses/InformationInSpeech/CDROM/Literature/LOTwinterschool2006/szabo.best.vwh.net/smart_contracts_2.html

Tapscott, D.: Getting Serious about Blockchain. McKinsey & Company, New York City (2017). https://www.mckinsey.com/industries/high-tech/our-insights/getting-serious-about-Blockchain

TelefonicaIoT: Adiós a los atascos| Welcome to The IoT World of Telefónica (2018). Iot.telefonica.com. https://iot.telefonica.com/blog/adios-a-los-atascos-0

The World Bank: Urban Development. World Bank (2018). https://www.worldbank.org/en/topic/urbandevelopment

The World Bank: Blockchain & Distributed Ledger Technology (DLT). World Bank (2018). https://www.worldbank.org/en/topic/financialsector/brief/blockchain-dlt

TomTom Traffic Index (2016). Tomtom.com. https://www.tomtom.com/en_gb/trafficindex/

UK Government Chief Scientific Adviser: Distributed Ledger Technology. Beyond Blockchain (2016). Assets.publishing.service.gov.uk. https://assets.publishing.service.gov.uk/government/uploads/system/uploads/attachment_data/file/492972/gs-16-1-distributed-ledger-technology.pdf

Vermesan, O.: Internet of Things strategic research roadmap. Internet of Things Strategic Research Roadmap, 9–52 (2009). http://internet-of-things-research.eu/pdf/IoT_Cluster_Strategic_Research_Agenda_2011.pdf

WEF: Top Ten Urban Innovations (2015). Www3.weforum.org. http://www3.weforum.org/docs/Top_10_Emerging_Urban_Innovations_report_2010_20.10.pdf

Wei, L.T., Yazdanifard, R.: The impact of Positive Reinforcement on Employees' Performance in Organizations. Am. J. Ind. Bus. Manage. **04**, 9–12 (2014)

WHO: Urban Transport and Health. Sustainable Transport. A Sourcebook for Policy-Maker in Developing Cities. Who.int (2011). http://www.who.int/hia/green_economy/giz_transport.pdf

Zanella, A., Bui, N., Castellani, A., Vangelista, L., Zorzi, M.: Internet of Things for smart cities. IEEE Internet Things J. **1**, 22–32 (2014)

Zielinski, S.: New Mobility: The Next Generation of Sustainable Urban Transportation (2012). Naefrontiers.org. https://www.naefrontiers.org/file.aspx?id=21915

A Critical Review of the Literature on Sectorial Analysis. Its Application to the Catalan Agricultural Sector

Elena Rondos-Casas(✉), Maria Angels Farreras-Noguer, and Salvador Linares-Mustarós

Business Department, University of Girona, Facultat de Ciències Econòmiques i Em, Campus Montilivi, 17071, Girona, Spain
{elena.rondos,angels.farreras, salvador.linares}@udg.edu

Abstract. Sectorial analysis adds qualitative aspects to business financial analysis and has traditionally been carried out using data aggregation. This work presents some of the current research lines in this field with the aim of identifying possible topics for future research. At the same time, we also study how the results can be applied to the agricultural sector in the Catalan economy and the quality of databases used to obtain related economic information.

Keywords: Sectorial analysis · Financial analysis · Agricultural sector

1 Introduction

The objective of financial analysis is to study the patrimonial, economic and financial situation of a company, determine its causes and, if necessary, identify any required corrections.

To this end, Calafell (1971) situated the analysis of financial documents within the deductive part of the integral methodological accounting process (PMCI) whereby, on the basis of verified - and consolidated if appropriate - financial documents, information is extracted regarding the status and development of the economic unit and subsequent conclusions are drawn for decision-making purposes.

In the same vein, Pirla (1970) defined financial analysis as the examination of accounting documents in order to assess the past, present and future situation of a company.

The analysis and interpretation of financial statements is used to study the solvency, profitability and/or liquidity of a company (or would these should be). The process involves obtaining initial information on financial statements and then selecting the important information or identifying which external aspects must be added to the analysis to make the information useful.

According to Rubio (2007), a set of technical and analytical instruments must be applied to these data to generate results that help the company's analysts or manager in their decision-making.

J. C. Ferrer-Comalat et al. (Eds.): MS-18 2018, AISC 894, pp. 83–96, 2020.
https://doi.org/10.1007/978-3-030-15413-4_7

According to Horrigan (1968), ratios have been one of the most widely used analytical techniques since the end of the 19th century. A ratio is a proportion that expresses the relationship between two numerical values, the idea being that it can provide more information than the individual figures themselves. On the basis of this, a diagnosis can be generated for a specific area of the company (Van Horne 2010).

A first approach (Westwick 1987; Ibarra 2006) established that it is necessary to compare these ratios with previous ones for the same company, with the standard established by the competitive context and with the ratios for the best and worst companies in the same sector. When adding qualitative aspects to the study, the use of sectorial data, that is to say, figures for other companies in the market, size or location, is key in complementing the analysis of financial statements (Oliveras and Moya 2005).

Sectorial analysis has historically been performed by calculating financial and economic ratios from the aggregation of companies' annual accounts (Amat 2018; Arimany 2016).

2 Methodology

The aim of this article is to review new lines of research in the field; it has been divided into five sections. First, we analyze research originating from limitations that may result from aggregating data and advances in the statistical tools used. A third line comprises authors who are concerned about the quality of the accounting information that some companies present. Finally, other works question whether it would be necessary to supplement such analysis with financial and non-financial indicators.

These lines of research allow us to identify new tools in sectorial analysis, while a further aim is to apply them to a specific business sector.

We chose the Catalan agricultural sector or one of its subsectors because of its weight in the Catalan economy. In 2017, the agriculture and fishing sector represented 1% of GVA (gross value added) in Catalonia, compared to 2.9% in Spain and 1.6% in the euro zone[1].. However, if we add the food industry, it becomes the main manufacturing sector in Catalan industry (19.77%). Furthermore, it is the second largest agri-food cluster in Europe and Catalonia's third largest export sector (Reguant 2018). It is worth highlighting that there is much investment in research in the sector, partly thanks to the IRTA (Institute of Research and Agri-food Technology), created in 1985 and dedicated to R+D+i (research, development and innovation). Because of the IRTA, there is cooperation between the private and public sectors, such as universities and research centers.

Within this sector we find some subsectors with interesting prospects for expansion, such as the winemaking industry, and others, such as dairy cattle rearing, that have great difficulties due to the free market and growing competition within the European Union (Reguant 2016).

We have analyzed different studies on the agricultural sector, the sectorial regulations and the type of database used to obtain information.

[1] IDESCAT web: https://www.idescat.cat/indicadors/?id=ue&n=10133&lang=&t=201700.

3 Lines of Research on Sectorial Analysis

3.1 Data Aggregation

Collecting economic and financial data at an individual level requires the use of descriptive statistics in order to obtain, describe, visualize and sum up the data and represent them in a numerical or graphic format. Obtaining data for a group or sector, on the other hand, might involve calculating the arithmetic average of data (Moya and Oltra 1993) or the weighted average of data for all companies studied (Montegut 2006).

These calculations may be used to reduce many numbers to a single digit, which is very useful but can at times fail to reflect reality. Therefore, it is recommendable to do further calculations. One example of this might be to introduce the concept of the liquidity return ratio (Linares et al. 2012), which obtains more reliable information on a sector by modifying its solvency capacity.

3.2 Statistical Tools

Other studies use statistical inference to generate predictions and models by taking into account uncertainty in the observations. This allows an overall assessment of companies by attending to different areas at the same time.

To this end, the following different methods and techniques can be used:

- Simple linear correlation and regression analysis. Correlation analysis is used to study whether two random variables are linked. On the basis of many observations, a correlation coefficient is calculated which ranges between +1 and −1. While 0 means no correlation, +1 displays a direct, strong correlation and −1 an indirect, strong correlation. Graphically, this can be represented in a dispersion diagram (Lind et al. 2008). A linear regression includes a dependent variable and an independent one. A correlation between these two variables is displayed by a straight line (Anderson et al. 2008).
- Multivariate analysis: this is a statistical method used to determine the contribution of more than one variable to an outcome, and it therefore includes multiple dependent variables and one independent variable (Anderson et al. 2008). This allows several ratios to be reduced to determine a single indicator for evaluating companies overall (Carranza et al. 2011).

Several techniques can be used in Multivariate analysis, including factorial analysis and discriminant analysis. Applications of discriminant analysis can be found in the famous work by Altman (1968) on predicting business failure and also in works evaluating improvements in financial indicators for companies in the food (Fontalbo et al. 2012) or coal (Fontalbo et al. 2012) sectors.

The aforementioned techniques or methods can use the whole population or samples, that is, statistical systems. We also find different criteria in the selection of company samples. While classical statistical methods employ a random sampling design, others advocate a non-random sampling, or a sample "based on a company's statement". This approach is used when the studied events happen to "few" companies in the sector, which therefore might go unnoticed in a random sample. However, it does not respect the population proportions of the sample, which could lead to overestimating the predictive capacity of the model (Palepu 1986).

Other mixed data collection methods are posited, some including random components and others not. We find examples of such systems in the prediction of business failure, where some authors apply random sampling (Balcaen and Ooghe 2006), others non-random sampling (statement-based sample) (Zmijewski 1984) and others a mix methodology (Mures et al. 2012).

3.3 Quality of Accounting Information

Sectorial analysis by aggregation is based on the individual annual accounts produced by companies in accordance with current regulations. Thus, accounting is the tool used to represent the economic reality of a company.

For a long time, no common criteria were available to allow this process to be performed in a homogeneous way and the same outcome obtained regardless of the subject matter.

One of the aims of accounting standardization was precisely this, to remove subjectivity from financial statements and make it possible to have comparisons over time and among several companies; that is, with each other and their environment.

The first Spanish law in this regard was Act 19/89, of 25 July, partially reforming and adapting Spanish commercial law to the ECC Directives on companies, which involved modifying the Commercial Code, applicable to all businesses.

The standardization process continued with approval of the 1989 Act on public limited companies and ended with the creation of the General Accounting Plan in 1990 (Royal Decree 1643/1990, of 20 December).

This plan was a very important instrument with regard to standardization, and was supplemented with all the sectorial accounting plans and resolutions by the Spanish Institute of Accountancy and Accounts Auditing (ICAC).

In the year 2000, changes were promoted to make the financial and economic data of European companies more comparable and consistent, regardless of place of residince or capital market. Thus, the European Commission requested that EU institutions required consolidated annual accounts to be formulated by applying the standards and interpretations issued by the International Accounting Standards Board (IASB), known as the IFRS (International Financial Reporting Standards).

European Parliament Regulation 1606/2002, relating to the EU's adoption of international accounting standards, gave renewed impetus to this process and a second phase of standardizing accounting data began.

This regulation specifies that any company producing consolidated annual accounts from 1 January 2005 had to apply IAS if its stocks were traded on the regulated market at the time of its balance sheet closing.

Act 62/2003, of 30 December, incorporated the international financial reporting standards (IFRS) adopted by the European Union for consolidated accounts into Spanish commercial law accounting. At the same time, Act 16/2007, partially reforming and adapting commercial legislation, introduced the changes necessary to move this process of international convergence forward into the Commercial Code and Companies Act. Finally, two legal texts approved in 2007 - the General Accountancy Plan (PGC)

(RD 1514/2007) and the General Accountancy Plan for Small and medium-sized enterprises, along with specific criteria for microenterprises (RD 1515/2007) – brought Spanish law into line with European Directives.

In accordance with the PGC, companies produce their financial statements every twelve months[2] and these are approved by the general meeting of shareholders (or partners) within a maximum period of 6 months. From that point on, this information can be accessed by the general public in the following way:

a. Listed companies must submit their annual accounts to the mercantile register and the CNMV (National Securities Market Commission). In these cases, companies must have a website with minimum content regulated by Circular 1/2004 of 17 March, including under the title "Information for shareholders and investors" all of the information required by Act 23/2003, of 18 July, such as its articles of association, annual report and documentation of general meetings, among others.
b. In general, companies must legalize their official books each year and deposit their annual accounts in the provincial Commercial Registry. These data are then organized and processed and transferred to the central Commercial Registry.

This regulation applies to public limited companies, limited companies, partnerships limited by shares, mutual guarantee companies, and, in general, any other company included in the legislation. It is also compulsory for foreign companies that have opened a branch in Spain. In addition, any individual businessperson who wishes to may also do so.

The Commercial Registry Regulation was approved on 19 July 1996, even though it was already undergoing reform due to the progress being made in related regulations. For example, Act 14/2013, of September 27, on entrepreneurs and their internationalization, establishes that registries "must be in electronic information format using a single computer system as determined by the regulation".

The information deposited in the Commercial Registry is public and can be consulted by anyone. Information can be obtained on paper or in a computerized format and is exportable on Excel spreadsheets through databases such as the SABI (Iberian Balance Analysis System), which has information on over 2.5 million Spanish and 700,000 Portuguese companies.

At present, it is possible to access it privately or via RedIRIS[3], which interconnects resources at universities and research centers.

Despite this being the age of information technology and access to information, different studies have expressed doubt that this information really presents a true picture regarding the assets, financial situation and results of some companies. This fact could pose a severe problem, since errors in these figures could lead to distortions in ratios and future studies.

The concept of creative accountancy was first defined by Naser (1993), who presented it as a process of manipulating the accounts to take advantage of legal gaps in the accounting regulations.

[2] Except in the cases of constitution, modification of the closing date or dissolution of the company.

[3] RedIRIS: Spanish Network interconnecting computer resources at universities and research centres.

Amat (2010) stated that creative accountancy is based on the following two axes:

- The use of various alternatives according to accounting standards. This makes it possible to use more or less conservative accounting criteria according to the interests of the company. The text includes examples, such as advancing or delaying a transaction, or changing accounting criteria.
- Applying more or less optimistic assumptions about future events. This has an effect, for example, when quantifying the financial statements and provisions.

Combining the two axes would explain how some companies embellish their annual accounts based on the regulations by increasing or reducing their assets, liabilities or profits.

Other authors, such as Ibarra (2006), for example, predict a perverse outcome from creative accounting since it serves to "delay and soften evidence of bad business phases, but does not guarantee their elimination and in the long term may lead to a price crash for companies with high stock prices".

Some studies provide specific examples of creative accounting and measures to combat it (Amat and Oliveras 2004).

Some current publications claim that there are companies that directly "lie" in their annual accounts (Amat 2017); that is, they make up the numbers, or in other words practice accounting fraud.

The same publications highlight early warning signs, ratios and calculations of formulas that should lead to suspicions that a company is consciously presenting wrong financial statements.

One solution could be only to analyze the financial statements of those companies that have been audited, but this could bias research at a sectorial level because not all companies are required to audit their annual accounts. According to Article 257 of the Royal Legislative Decree 1/2010, of July 2, the revised text of the law on capital companies, those companies are required to be audited that meet two of the three following requirements for two consecutive financial years on the closing date of each:

- Its total assets exceed €2.85 million
- Its net turnover exceeds €5.7 million
- Its average number of employees during the financial year exceeds 50.

It is also compulsory for the following[4]:

- Companies that receive subsidies or grants (from public administrations or the European Union) for an accumulated amount of €600,000.
- Companies that have contracts with the public administration that exceed €600,000 and represent more than 50% of their turnover.
- Some foundations, credit unions, and companies that act as financial intermediaries, issue securities accepted for trading on regulated markets or multilateral trading systems, companies that issue public bonds and companies subject to private insurance regulations.

[4] According to Royal Decree 1517/2011 on Audit Regulations.

- Any other company where it is established in the statutes, or at the request of the Commercial Registry, or by court order.

Despite the above, data from the Central Directory of Companies (DIRCE)[5] reveal that there were 3.28 million active enterprises in Spain in 2016, while only 59,598 were audited according to the Spanish Institute of Chartered Accountants (ICJCE); that is, only 1.82% of the total.

3.4 Financial Indicators

When a business ratio is analyzed it is often compared to its optimal value according to that defined by different authors (Amat 2008; Álvarez 1983).

Other authors compare the ratios of the company with ideal ratios (Amat and Fiestas 2000), which according to Amat (2018) are "the average value of a certain ratio for a group of companies characterized by its good financial economic results, normally belonging to a particular sector of the economy".

Therefore, a sectorial study can only be done taking into account those companies that can be considered leaders in the sector. This select group provides us with a guide for performance.

Some studies claim that this small group should always comprise high-growth companies, that is, those which have maintained an average annual growth of over 20% for three consecutive years. This growth can be measured in employees (based on a minimum of 10 for the first year analyzed) or by business volume (Eurostat-OCDE 2008). Another possibility could be the subgroup comprising the so-called gazelle companies (Amat et al. 2010), which are high-growth companies that have been in existence for less than five years.

Other authors criticize the idea of establishing a single value for a ratio on the basis that the accounting data are not always precise but open to subjectivity. According to Reig, Sansalvador and Trigueros (Reig et al. 2010) it is necessary "to leave aside the binary logic of Aristotle and incorporate fuzzy logic into situations of uncertainty and subjectivity".

The concept of fuzzy numbers (Zadeh 1975) has spread to many scientific disciplines and constitutes a way of treating subjectivity and/or uncertainty in various fields, including business problem solving.

Triangular fuzzy numbers (Kaufmann and Aluja 1986) have been used in models to predict marketing and logistics issues or in financial analysis (Lafuente 2001).

It could also be useful to move from several ratios explaining the situation of a company or sector to one single indicator that contributes an overall vision. This technique was first applied by Altman (1968), who used the Z-score formula to measure the probability that a company will have insolvency problems. The Z-index (Vladu et al. 2017) has aso been used to create a model to identify companies with a high probability of having made-up accounts.

[5] https://www.ine.es/prensa/dirce_2017.pdf.

Finally, Vázquez, Guerra and Tellez, (Vázquez et al. 2011) posited a ratio that allows companies to be classified and an overall assessment made for each one. This can be used in tandem with the traditional technique employing Multivariate models.

3.5 Non-financial Indicators

When analyzing companies or business sectors, companies' internal and external financial data are taken into account. According to Horngren, Foster and Datar (Horngren et al. 2002), it is necessary to also include internal non-financial data, such as level of sales per employee, time of attention given to customers, and motivation in the workplace, and external non-financial data, such as level of customer satisfaction or decrease of sales by distributor, among others.

These types of indicators can facilitate better decision-making provided they are not overly complicated and company workers are willing to participate. Otherwise, they could have the opposite effect (Morillo 2004).

Such non-financial data might also include other concepts such as quality management or corporate social responsibility. Thus, we have moved on from the declarations made by Friedman (1970), who claimed that the only responsibility of a company is to make profits, to the idea that companies must have "Corporate Social Responsibility" (CSR). This concept is based on Caroll's (1979) idea that companies' operations must be beneficial on an economic, social and environmental scale.

With regard to the cost involved, there are authors who believe that "Social responsibility should be considered an investment and not an expense" (CCE 2001) "in the same way as quality is".

Another aspect currently under investigation is whether Corporate Social Responsibility can benefit the financial profitability of a company or not. While some studies have found a positive correlation in this respect (Miras et al. 2014), others have not. Madorran and García (2016) attempted to relate the financial results of Spanish companies listed on the IBEX from 2003 to 2010 to Corporate Social Responsibility and were not able to establish a clear relationship.

4 Application to the Agricultural Sector

4.1 Research on the Agricultural Sector

Economic studies on the agricultural sector have always been numerous and can generally be divided into two types: those devoted to analyzing business failure in the sector and those dealing with the use of accounting.

While financial problems suffered by farms had appeared earlier, from the 1970's onwards some articles were published in the United States using discriminant models to study the use of loans by farming families (Krause and Williams 1971).

In the 1980's, we find studies on the effects of the crisis on the sector (Murdock and Leistriz 1988) and on non-viable agricultural holdings.

The early 1990's saw the emergence of studies using logit regression models to classify farms according to their financial situation (Wadswoth and Bravo-Ureta 1992), while other authors studied farmer insolvency and related it to the price of land and common agrarian policies (CAP) (Davis 1996). Fieldwork studies were also done on failed companies and abandoned fields in the United States (Stam et al. 1899) and Estonia (Lukason 2014).

It is also worth highlighting the research conducted on factors that positively affect the value of farming operations, such as population or the possibility of obtaining credit, and others that negatively affect it, such as indebtedness (Devadoss and Manchu 2007).

Another line of research in the agricultural sector has been to determine whether accounting is useful for analyzing the sector or not. While Poppe (Poppe 1991) pointed out that accounting information is not widely used despite it benefitting farms, Argilés (2001) argued that accounting regulations need to be more developed in the sector to benefit farm management; some fieldwork has been done in this regard (Argilés and Slof 2003).

This idea has continued to be evaluated, even after implementation of the NIC41 (Vazakidis et al. 2010). That said, its use is still in its early stages in some countries (Dutta and Das 2017).

4.2 Regulation of the Sector

Regarding specific regulations for the sector, a French agricultural accounting plan was published in 1970, although the first compulsory one came into effect in 1987. It was a model that allowed economic-financial information to be standardized with other sectors of activity in terms of quantity and quality and at the same time attempted to provide solutions to topics specific to the sector, such as biological assets, self-consumption, compensation and subsidies. We also find some other experiences at an international level in Australia or New Zealand, the latter with Management Accounting for the New Zealand Farmer (1977).

In Spain, approval of the General Accounting Plan in 1990 (RD 1643/1990) meant the standardization of general accounting methods in the country. Specific to the sector, a sectorial plan applicable to companies in the wine sector was adopted in 2001 (Order 11 May, 2001).

From an international perspective, it was not until the year 2000 that the International Accounting Standards Committee (IASC) adopted the final version of the International Accounting Standard on agriculture, IAS41, which has basic applications on topics such as:

- Biological assets (valued at fair value less estimated costs as long as they can be determined with sufficient reliability).
- Agricultural products during harvest.
- Official subsidies received by biological assets except where IAS 20 applies (Accounting for government grants and disclosure of government assistance).

These regulations came into force from 2003 and were mandatory as of 2005 for listed Spanish companies that present consolidated financial statements.

According to Arimany, et al. (2013), many agricultural companies do not apply the concept of reasonable value as they believe it cannot be determined reliably and only 31% of companies in the study followed the provisions with regard to subsidies.

Recently, Argilés et al. (2018) have shown that using fair value and not the historical cost of biological assets helps anticipate cash flows better, thus improving the quality of accounting information for these companies.

The rest of the companies in the sector have been obliged to apply general accounting standards according to the type of enterprise (micro, small, medium-sized or large company) since 2008.

4.3 Agricultural Sector Data

At the agricultural sector level, data can be found in statistical compilations from different yearbooks and reports. We find data in IDESCAT (Statistics Institute of Catalonia) and INE (National Institute of Statistics), External Trade reports (ICEX) and statistics from the ecological sector in Catalonia on the website of the Catalan Council of Ecological Agrarian Production.

While companies must submit their annual accounts to the Commercial Registry or the CNMV, in the agricultural sector there are still many economic operators who work under the tax system for individual entrepreneurs, civil societies or associations (who do not need to submit annual accounts if they do not want to) or cooperatives in the General Cooperatives Regsistry.

This means that accounting data for companies in the agricultural sector from the SABI database are mostly for limited companies (SL) or public limited companies (S.A.).

A further source of information is RICA (the Network of Farm Accounting Data), which was created by the Council of the European Economic Community (EEC) in 1965 through regulation 79/65. Its main function is to measure farming income in Europe and provide economic analysis. It is a European system of voluntary sampling surveys that collect annual data on 800,000 farms (of about 5 million in total). Its objectives are to analyze the income and economic activities of farms and assess the impacts of the Common Agrarian Policies (CAP). Recent studies have investigated whether the data it includes should be expanded with other information such as innovation, age of assets or farm insurance (Vrolijk et al. 2016).

The data RICA works with are collected at the accounting offices of each country. Although Spain has the National Agricultural Accounting Network (RECAN), the offices that collect the data are located in the individual autonomous regions. Thus, in Catalonia we have the Agricultural Accounting Network of Catalonia (XCAC), which presents these data free and in PDF and Excel format on its website in order to allow for the corresponding studies to be conducted. The last published data (2016) study 597 Catalan farms of the 25,968 counted, according to the network.

5 Conclusions and Future Research Lines

Financial statement analysis using economic-financial ratios can be traced back to the last half of the 19th century. Over time, ratios have evolved and analysis has improved in quality when comparing individual data with a group of similar companies. This is the origin of sectorial analysis, which has traditionally been calculated by aggregating accounting data. Unfortunately, this system has some limitations, which is why new lines of research have originated.

With regard to statistical tools, sectorial ratios calculated as a mean or average are very useful but may not reflect the true reality. Providing variations in how they are calculated, such as the sectorial liquidity return ratio, can provide more realistic data.

If sectorial analysis is performed on the basis of statistical inference, it will be necessary to determine whether a simple linear correlation or a Multivariate method is needed and how samples are to be obtained. The approach of using a single ratio that explains the overall situation of a company or a sector is also an issue to consider in the future.

With regard to accounting information, having accounting statements that reflect a true picture of the companies in a sector is essential if research is to be rigorous. In addition, ratios based on the specific characteristics of the sector could be calculated in order to provide more relevant information for the sector analyzed.

On the business level and in terms of financial indicators, the ratio is compared with an "optimal value" defined by different authors. In sectorial analysis, "ideal ratios" can be calculated using the values of leading companies. It is also possible to consider calculating the "minimal values" that guarantee permanence in a sector.

Finally, non-financial indicators might provide key information to properly classify the companies within a sector, although more studies are needed to confirm their effectiveness.

Advances in each line of research could be applied to the agricultural sector due to its unique character. As a whole, it might be worth analyzing both the minimum values that ensure the continuity of companies and application of the industry's own nomenclature. A further possibility would be to analyze agricultural sub-sectors, such as winemaking, which has the highest prospects of further expansion.

These proposals could lead to better sectorial analyses, which would in turn help companies to make better decisions.

References

Altman, E.: Financial ratios, discriminant analysis and the prediction of corporate bankruptcy. J. Finance **23**(4), 589–609 (1968)

López, J.Á.: Análisis de balances. Editorial Donostiarra, S.A. (1983)

Amat, O.; Fiestas, I.: Ratios de las empresas más rentables, Ed. Gestión 2000, Barcelona (2000)

Amat, O.; Oliveras, E.: Propuestas para combatir la contabilidad creativa. Universia Business Review-Actualidad Económica - Primer Trimestre (2004)

Amat, O.: Fiabilidad de la nueva normativa contable (PGC y NIIF) y Detección de Maquillajes Contables. Boletín de Estudios Económicos **65**(199), 93–104 (2010)

Amat, O.; Fontrodona, J.; Hernández, J.M.; Stoyanova, A.: Las empresas de alto crecimiento y las gacelas (primera ed.), Ed. Profit, Barcelona (2010)
Amat, O.: Análisis de balanços, Ed. Profit, Barcelona (2015)
Amat, O.: Empresas que mienten: como maquillan las cuentas y como detectarlo a tiempo, Ed. Profit, Barcelona (2017)
Amat, O.: Ratios Sectoriales 2016. Cuentas nuales (balances y cuentas de resultados de 166 sectores) 25 ratios para cada sector. Ed. ACCID, Barcelona, (coordinador) (2018)
Amat, O.: Anàlisis integral d'empreses, Ed. Profit, Barcelona (2018)
Anderson, D.; Sweeney, D.; Williams, T.: Estadística para administración y economía (10a ed.), Ed. Prentice Hall, México (2008)
Argilés, J.M.: Accounting information and the prediction of farm non-viability. Eur. Account. Rev. **10**(1), 73–105 (2001)
Argilés, J.M., Slof, E.J.: The use of financial accounting information and firm performance: an empirical quantification for farms. J. Account. Bus. Res. **33**(4), 2003 (2003)
Argilés-Bosch, J.M., Miarons, M., Garcia-Blandon, J., Benavente, C., Ravenda, D.: Usefulness of fair valuation of biological assets for cash flow prediction. Span. J. Financ. Account./ Revista Española de Financiación y Contabilidad **47**(2), 157–180 (2018). https://doi.org/10.1080/02102412.2017.1389549
Arimany-Serrat, N., Noguer, M.À.F., Tarrés, J.R.: Alejados de la NIC41: ¿Es correcta la valoración del patrimonio neto de las empresas agrarias? Economía Agraria y recursos naturales **13**(1), 27–50 (2013)
Arimany-Serrat, N., Noguer, M.À.F., Tarrés, J.R.: Economic and financial analysis of rioja wine sector. Intang. Cap. **12**(1), 268–294 (2016)
Balcaen, S., Ooghe, H.: 35 years of studies on business failure: An overview of the classical statistical methodologies and their related problems. Br. Account. Rev. **38**(1), 63–93 (2006)
Calafell Castelló, A.,Apuntes de Introducción a la Contabilidad, Departamento de Contabilidad de la Facultad de Ciencias Económicas y Empresariales, Universidad Autónoma, Madrid, Curso, 1971–1972 (1970)
Carroll, A.B.: A three-dimensional conceptual model of corporate performance. Acad. Manag. Rev. **4**, 497–505 (1979)
Comisión de las Comunidades Europeas (2001): COM(2001) 366 final. Libroverde Fomentar un marco europeo para la responsabilidad social de las empresas. Bruselas, 18 September 2001
Davis, A.S.: Insolvency in agriculture: bad managers or the common agriculturalpolicy? J. Appl. Econ. **28**(2), 185–193 (1996)
Devados, S., Manchu, V.: A comprehensive analysis of farm land value determination: a county-level analysis. J. Appl. Econ. **39**(18), 2323–2330 (2007)
Dutta, H., Das, D.: Accounting for farms in India: an analysis in the context of recognition, measurement and presentation of financial data. IUP J. Account. Res. Audit Pract. Hyderabad Tomo **16**(3), 34–53 (2017)
Pirla, J.M.F.: Teoría económica de la contabilidad. Introducción contable al estudio de la economía, Ed. ICE, Madrid (1970)
Fontalvo Herrera, T.; De La Hoz Granadillo, E.; Vergara, J.C.: Aplicación de análisis discriminante para evaluar el mejoramiento de los indicadores financieros en las empresas del sector alimento de Barranquilla-Colombia. Ingeniare, **20**(3) (2012)
Herrera, T.F., Gomez, J.M., Granadillo, E.D.L.H.: Aplicação da análise discriminante para avaliar o desempenho dos indicadores financeirosnas empresas do setor de carvãona. Colômbia **8**(2), 64–73 (2012)
Friedman, M.: The social responsibility of business is to increase its profits. N. Y. Times Mag. **32–33**, 122–124 (1970)

Lafuente, A.M.G.: Nuevas estrategias para el análisis financiero en la empresa. Ariel Economía, Barcelona (2001)
Guide comptable professional des entreprises agrícoles. I.G.E.R., Paris (1970)
Guide comptable professional des entreprises agrícoles. I.G.E.R., Paris, (1987)
Horrigan, J.: A short history of financial ratio analysis. Account. Rev. **43**(2), 284–294 (1968)
Horngren, Ch., Foster, G., Datar, S.: Contabilidad de Costos: Un enfoque gerencial. Ed. Prentice Hall, México (2002)
Ibarra Mares, A.: Una perspectiva sobre la evolución en la utilización de las razones financieras o ratios. Pensamiento y Gestión **21**(1), 234–271 (2006)
Instituto Censores Jurados de Cuentas de España (ICJCE) Fotografía del sector auditor en España. Estimación de su impacto económico 2006–2016
Kaufmann, A., Aluja, J.G.: Introducción de la teoría de los subconjuntos borrosos a la gestión de las empresas. Ed. Milladoiro. Santiago de Compostela (1986)
Krause, K.R., Williams, P.L.: Personality characteristics and successful use of credit by farm families. Am. J. Agric. Econ. **53**(4), 619–624 (1971). https://doi.org/10.2307/1237826
Lind, D., Marchal, W.G., Watlen, S.A.: Regresión lineal y correlación en Estadística aplicada a los negocios y la economía. 13th° edn., pp. 458– 496. McGrawHill, México (2008)
Linares, S., Farreras, M.A., Ferrer, J.C., Rabaseda, J.: Una nueva ratio sectorial. La ratio de retorno líquido. Cuadernos del Cimbage, **15**, 57–72 (2012)
Lukason, O.: Why and how agricultural firms fail, Evidence from Estonia. Bulg. J. Agric. Sci. **20**(1), 5–11 (2014)
Madorran, C., Garcia, T.: Corporate social responsibility and financial performance: The Spanish Case. Rev. Adm. Empres **56**(1), 20–28 (2016)
Management Accounting for the New Zealand Farmer (NZSA 1977)
Miras Rodríguez, M.M., Carrasco Gallego, A., Escobar Pérez, B.: Corporate social responsibility and financial performance: a meta-analysis. Span. J. Financ. Account./Revista Española de Financiación y Contabilidad **43**(2), 193–215 (2014)
Salla, Y.M.: Análisis de los factores explicativos del éxito competitivo en las almazaras cooperativas catalanas. Tesis doctoral, Universitat de Lleida, Lleida (2006)
Morillo M.: Indicadores No financieros de la contabilidad de gestión: herramienta del control estratégico. Actualidad Contable Faces, Año **7**(8), 70–84, Enero-Junio 2004
Moya, I., Oltra, M.J.: Las empresas agroalimentarias. Un análisis empresarial y bursátil, CIRIEC-España, Revista de Economía Pública, Social y Cooperativa **15**, 207–238 (1993)
Murdock, S.H.; Leistritz, F.L.: The farm financial crisis: Socioeconomic dimensions and implications for producers and rural areas. Westview Pr. (1988)
Quintana, M.J.M., Gallego, A.G., Pascual, M.E.V.: Análisis del fracaso empresarial por sectores: factores diferenciadores, Pecunia: revista de la Facultad de Ciencias Económicas y Empresariales, Nº. 1, 53–83 (jul–dic 2012). ISSN 1699-9495
Naser, K.: Creative Financial Accounting: its nature and use, Ed. Prentice Hall, Londres (1993)
Oliveras, E; Moya, S.: La utilización de los datos sectoriales para complementar el análisis de los estados financieros Revista de Contabilidad y Dirección **2**, año 2005, pp. 53–69 (2015)
Palepu, K.G.: Predicting takeover targets: a methodological and empirical analysis. J. of Account. Econ. **8**(1), 3–35 (1986)
Poppe, K.J.: Information needs and accounting in agriculture. Agric. Econ. Res. Inst. LEI **444**, 1–51 (1991)
Reguant, F.: El sector català de vaquí de llet front al mercat global. Observatori Economia Agroalimentària -Collegi Economistes de Catalunya (2016). https://obealimentaria.wordpress.com/2016/06/06/28/
Reguant, F.: L'agroalimentació del segle XXI a l'economia catalana, ObservatoriEconomiaA-groalimentària - CollegiEconomistes de Catalunya (2018)

Mullor, J.R., Selles, S., Enrique, M., Pin, T., Antonio, J.: Lógica borrosa y su aplicación a la contabilidad. Revista Española de Financiación y Contabilidad **29**(103), 83–106 (2000)
Rubio D.P.: Manual Análisis Financiero. España. Universidad de Málaga. Edición electrónica (2007). www.eumed.net/libros/2007a/255
Stam, J.M., Dixon, B.L.: Farmer bankruptcies and farm exits in the United States, 1899–2002. United States Department of Agriculture, Washington, 36 p. (2004)
Wadswoth, J.J., Bravo-Ureta, B.E.: Financial performance of New England dairy farms. Agribus. Int. J. **8**(1), 47–56 (1992)
Van Horne, J., Wachowicz, Jr., J.M.: Análisis de los estados financieros. Fundamentos de Administración Financiera, pp. 136–138. Pearson, Mexico (2010)
Vazakidis, A., Stergios, A., Laskaridou, E.C.: The importance of information through accounting practice in agricultural sector-european data network. J. Soc. Sci. **6**(2), 221–228 (2010)
Carrazana, X.V., Fonseca, A.G., Tellez, I.A.: Aplicación de métodos multivariados: una respuesta a las limitaciones de los ratios financieros", en Contribuciones a la Economía, nº 2011–01 (2011)
Vladu, A.B., Amat, O., Cuzdrorien, D.D.: Truthfulness in Accounting: How to discriminate accounting manipulators from non-manipulators. J. Bus. Ethics **140**(4), 633–648 (2017)
Westwick, C.A.: Manual para la aplicación de los ratios de gestión, Ed. Deusto, España (1987)
Vrolijk, H., Poppe, K., Keszthelyi, S.: Collecting sustainability data in different organisational settings of the European Farm Accountancy Data Network. Stud. Agric. Econ. **118**, 138–144 (2016)
Zadeh, L.A.: Fuzzy sets. Information and control **8**(3), 338–353 (1965)
Zmijewski, M.E.: Methodological issues related to the estimation of financial distress prediction models. J. Account. Res. Suppl. 22 (1984). Studies on Current Econometric Issues

Economic Financial Health of the Wine Sector in Time of Financial Crisis: Catalonia, La Rioja and Languedoc-Roussillon

Núria Arimany-Serrat[1] and Maria Àngels Farreras-Noguer[2](✉)

[1] Department of Business Administration, University of Vic-UCC, Sagrada Familia n. 7, 08500 Vic, Barcelona, Spain
nuria.arimay@uvic.cat

[2] Department of Business Administration, University of Girona, Campus Montilivi, carrer de Girona 10, 17071 Girona, Spain
angels.farreras@udg.edu

Abstract. The wine cluster has an outstanding presence in the food industry and in this study of three zones of winery tradition: Catalonia, La Rioja and Languedoc-Roussillon, where an economic financial analysis and the correspondent comparative in the difficult period of the 2008–2013 crisis is the intended focus.

To make a comparative of the analysis of the economic and financial situation of these wine companies in the period 2008–2013, the main financial indicators were compared using the appropriate descriptive statistics for the three companies studied.

The crisis period analysed is a period of changes in the financial economic health of wine companies due to the effects of innovation, internationalization and crisis in the three zones. In general, being large companies, in this period, the exporting tradition favours its economic financial health, and, despite being mostly family businesses, they are companies with differentials in products and processes.

Keywords: Wine sector · Economic financial analysis · Sector accounting

1 Introduction

The wine cluster has an outstanding presence in the food industry and in this study of three zones of winery tradition: Catalonia, La Rioja and Languedoc-Roussillon, where an economic financial analysis and the correspondent comparative in the difficult period of the 2008–2013 crisis is the intended focus. In other words, the objective of the research is to determine the economic financial health of the large companies which produce wine in Catalonia, La Rioja and Languedoc-Roussillon in the period 2008–2013, using financial indicators endorsed by the literature. The methodology used sets off from the accounting data gathered through the database Iberian Balance Analysis System (SABI), and the AMADEUS database (European database) of one sample from large Catalan companies, La Rioja and Languedoc-Roussillon, wine producers who have a minimum operating revenue of 8.000.000€ and minimum assets of 4.000.000€

J. C. Ferrer-Comalat et al. (Eds.): MS-18 2018, AISC 894, pp. 97–111, 2020.
https://doi.org/10.1007/978-3-030-15413-4_8

for two consecutive years. To subsequently calculate the financial indicators and combined with descriptive statistics to carry out the conventional analysis of the financial states, by adding a specific analysis to the equity variations, in order to obtain results and conclusions which allow these large wine companies to set market positions.

Revised the literature in regard to the economic financial analysis of the wine companies of the three zones and companies of the study, the research has been structured into four distinct parts: short-term analysis of the financial situation, long-term analysis of the financial situation, economic analysis, and equity analysis. The analysis of the short-term financial situation will allow us to assess the capacity of these companies to meet their short-term payment obligations, the analysis of the long-term financial situation will allow us to measure the capacity of the company to meet long-term debts, and the economic analysis or of results will show how the results have been produced, identifying the reasons triggered during/in the economic situation in the period analysed. Then, the equity analysis will allow us to identify if the companies are being capitalized, offering more guarantees to third parties directly related to them.

Finally, at the conclusions, a diagnosis is performed about the financial economic situation of the companies of the three zones analysed, and the corresponding comparative in the period 2008–2013 under study.

2 Preliminary

Catalonia has a long tradition in the wine industry and has recognized twelve "Denominacions d'Origen (DO's)" or Appellations of Origen: DO Penedès, DO Terra Alta, DO Catalonia, DO Tarragona, DO Conca de Barberà, DO Costers del Segre, DO Empordà, DO Montsant, "Denominació d'Origen Qualificada" or qualified Appellation of Origen (DQQ) Priorat, DO Alella, and DO Pla de Bages. On the other hand, two thirds of sales are destined for exportation.

La Rioja is also an area with a long tradition in the wine industry, and in 1925 obtained the first DO granted in Spain, and in 1991 had DOC, in Spanish "Denominación de Origen Cualificada". The wine cluster in La Rioja generates an important added value to the region and creates many jobs, in addition, a third of the wine is exported to the main world markets, evidencing the competitiveness of the sector, with a positive balance of foreign trade, and with outstanding effects on the society and economy of La Rioja.

The Languedoc-Roussillon area has experienced major changes in the economic field, as it has had to make a firm commitment to innovation and quality within the framework of the wine companies, in order to face greater competition at the global level, taking into account the environmental obligations and the internationalization of the sector itself. The diversity of the region accounts for 40% of the French vineyard by the soils and climates which make up the different denominations, so it has a specific weight in the French wine sector, it must be remembered that the Languedoc-Roussillon has more than 30 wines which have achieved the category AOC, namely, the appellation of origin which in France is called Appellation d'Origine Contrôlée (AOC). Identified this outstanding presence in this French area of Languedoc-Roussellon, it is important to assess the financial economic health of its wine companies in the crisis period 2008–2013, in order to be able to carry out the comparison with the other areas of the study and to focus on innovation and internationalization of this particular sector in the period of systemic crisis.

3 Review of the Literature

There is a variety of academic literature in reference to the wine industry, especially in the five main wine-producing countries: France, Italy, Spain, Germany and Portugal (Roman-Sánchez et al. 2015), although in an economic financial health level there is only a few academic publications focusing on these crisis periods; a publication referring to the wine industry in the Languedoc Roussillon identifies that a solution to the crisis of the wine industry is concretized in the competitive strategies of specialization and differentiation with respect to the low cost strategies (Dukesneois et al. 2010), but there are no publications about financial economic health through benchmark financial indicators for the period 2008–2013 in the three areas under study.

In the wine sector, mercantile companies and, also cooperatives, have different economic and financial structures, in particular, cooperatives have a smaller economic structure, are less solvent and have more indebtedness than mercantile companies (Gómez-Limón et al. 2003). On the other hand, the wine industry has been affected by the changes of European regulation, reason why the capitalist companies obtain better returns than the cooperatives, caused by greater internationalization, better commercial strategy as a result of the corporate structure, the size, and financial structure. Also, it should be kept in mind that cooperatives are more focused on the sale of bulk wines (Castillo-Valero and Garcia-Cortijo 2013). Thus, the concentration of these companies favours internationalization, for access into more resources which facilitate exports (Carneiro et al. 2014). And, apart from size, other factors favouring exports are professional experience, geographical location and institutional support (Bardají et al. 2014).

Increasingly strong international competitiveness in sectors such as the food industry highlights the fact that smaller firms find it difficult to access international markets (Barmen 2014), as, the greater concentration, the greater investment in marketing, gives an increase in collective reputation and appellations of origin (Marchini et al. 2015). Even so, small producers have a place in the wine world, as many have differentiated themselves into product types, and attract much of the oenological tourism (Francioni et al. 2016). In addition, in the case of peri-urban regions these wine companies have adapted their business models respecting the social benefits inherent to them (Recasens et al. 2016). And, from the Spanish perspective, it should be borne in mind that winemakers consider that a good product is offered which clearly identifies its origin (with appellations of origin which generate sustainable competitive advantages) with wines of quality and territorial differentiation (Garcia-Galan 2014).

Regarding financial economics health, recent studies identify that the variable family is relevant to achieve good economic and financial profitability in both France and Italy (Bresciani et al. 2016), with financial differentials between the family companies and non-family ones (Broccardo et al. 2015), also the local implications influences on business success, as endorsed by studies on cluster-based wine profitability in Chile and Italy (Giuliani 2013), and knowing the customer and the brand significantly favours trade (Thomas et al. 2013). On the other hand, economic financial health is directly related to the innovation of these companies, since technological changes affect productivity and results (Sellers-Rubio et al. 2016) and, also cutting-edge technologies along with the environmental regulations allows a more sustainable production of wine (Roman-Sánchez et al. 2015). It is also important to make evident that when innovation and tradition interact, it provides the desired competitive advantage. Innovative processes and products should be in tune with tradition and passion for wine, especially in family wine-growing enterprises where it's necessary to innovate at the same time as spreading the family business philosophy for future generations (Vrontis et al. 2016). With respect to gender, nowadays, women are more present in the wine industry, but still it's a long way until full parity on executive positions, or to participate in board membership of these large companies, endorsed by the Australian study in the period 2007–2013 in a representative sample of companies in the wine sector (Galbrealh 2015).

In recent years, these large wine companies, familiar and non-familiar ones, have experimented structural and economic changes, and in the crisis period under study 2008–2013, it is desired to identify financial economic health of three areas highlighted in wine production, in order to carry out the opportune comparative with the aim of analysing the sector currently present in the international markets.

4 Comparisons of the Economic Financial Analysis of the Catalan Wine Sector, La Rioja and Languedoc-Roussillon

In this study, the financial economic analysis of Catalan wine sector (formed by a sample of 14 companies located in Cataluña) is compared with the wine sector of La Rioja (formed by a sample of 16 companies) and the wine sector of Languedoc Roussillon (formed by a sample of 16 companies). Most of the companies are large (due to the selection criteria of the three considered samples) and, they are companies which show normal annual accounts, according to what the General Accounting Plan states.

The samples have been obtained through the data provided in the database SABI (as for Spain) and AMADEUS (as for France) with the following criteria:

Main economic activity: Wine production according to the Code CNAE 2009: 1102.
State: Active.
Minimum operating revenue: 8.000.000€.
Minimum assets: 4.000.000€.

To make a comparative of the analysis of the economic and financial situation of these wine companies in the period 2008–2013, the main financial indicators were compared using the appropriate descriptive statistics for the three companies studied.

First, a comparative of the analysis in the short-term financial situation was performed, followed by long-term financial analysis and result analysis, also considering the changes in net worth, in order to assess the capitalization of these companies.

To perform this financial economic analysis, the indicators analysed have been:

1. For the short-term analysis: the working capital (current asset - current liabilities) and short-term solvency (current assets ÷ current liabilities).
2. For the long-term analysis: indebtedness (total debts ÷ net worth and liability); The quality of the debt (short term debts ÷ total debts) and the asset turnover (operating revenues ÷ total assets).
3. For the analysis of results: return on equity (ROE = Net Result ÷ Net equity - Net result); return on assets (ROA = EBIT ÷ Asset); Value added (operating revenue – operating expenses) and staff productivity (value added ÷ personnel costs).
4. For capitalization: the evolution of net worth in the analysed period is calculated by the corresponding horizontal percentage.

4.1 Comparative of the Analysis of the Short-Term Financial Situation

The main objective of the analysis of the short-term financial economic situation is to determine the capacity that these wine companies have, in order to meet their short-term payment obligations.

First, the median of the equity components and distribution of themselves for the three analysed samples:

The structure of the distribution of equity components is similar, although there is a difference in the percentages that make up non-current assets and current assets. As shown (Fig. 1), the non-current and current assets of the Catalan wine sector, using medians, are at 50% between non-current and current assets, but in the case of La Rioja (33.26%), and Languedoc-Roussillon (24.33%), the average of non-current assets shows a lower percentage so that current assets dominate predominantly the total assets of the Rioja and Languedoc Roussillon (especially) wine companies, namely, they are companies, which, in median values, have not as many investments in non-current assets as Catalan companies. However, there is an evolution on these equity components, clearly differentiated between the three zones during the period between 2008–2013. In this sense, both in the Catalan wine companies and in the wine companies of La Rioja, a tendency of declination is shown: highlighted in the first ones, the increase in non-current assets by 3% and the decrease in net worth by 4%, which demonstrates the slight decapitalization of these Catalan companies and, in the second ones, the reduction in non-current liabilities by 50% and in current liabilities by 39% is relevant. On the other hand, in the Languedoc-Roussillon wine companies there is a growing tendency of the equity components, emphasizing the increase of non-current liabilities by 116% and the increase in non-current assets by 75%.

Then, comparing the working capital, we can see that in the three zones it is positive along the six years and the evolution is inverse; namely, while in the Catalan wine sector the working capital go down in the period 2008–2013, in the wine sector of the Rioja and Languedoc-Roussillon it increases considerably (Fig. 2).

Regarding the short-term solvency ratio, which measures the company's capacity to cope with short-term debt by carrying out its current assets, in the six years and in the three samples, companies show a good capacity to meet commitments in the short term. In the case of Catalan wine companies, short-term solvency is acceptable in the period

2008–2013 (despite the economic downturn in 2011); in the case of the Languedoc-Roussillon it is acceptable, and its evolution keeps increasing throughout all the periods until reaching its maximum value in the year 2013. And, in the case of the wine companies of La Rioja they even have idle current assets, which must be invested correctly.

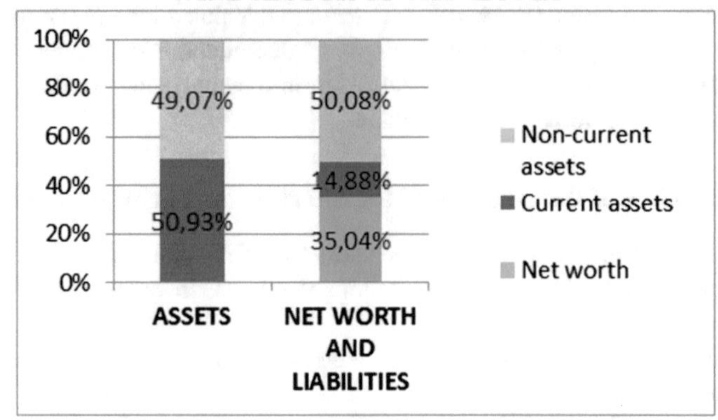

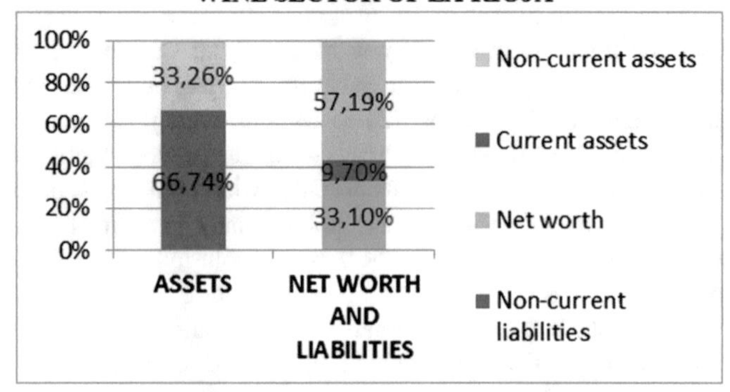

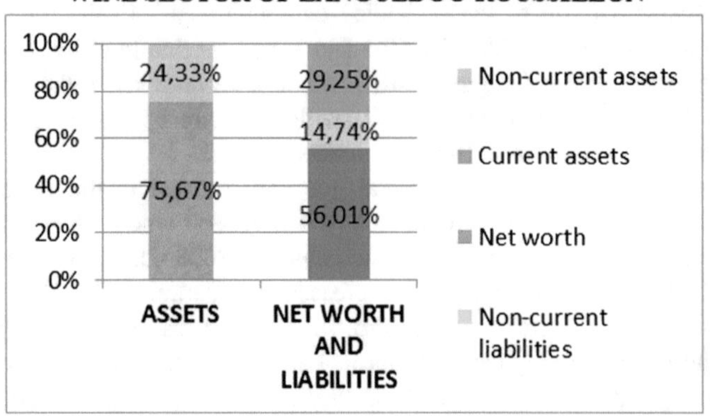

Fig. 1. Median of the equity components of the balance sheet of the three areas analysed in the period 2008–2013

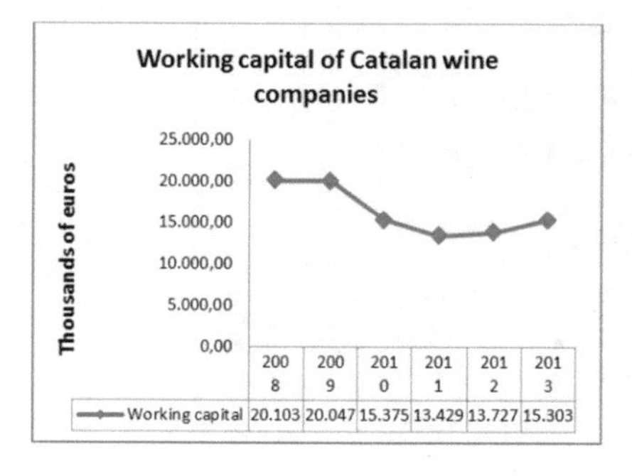

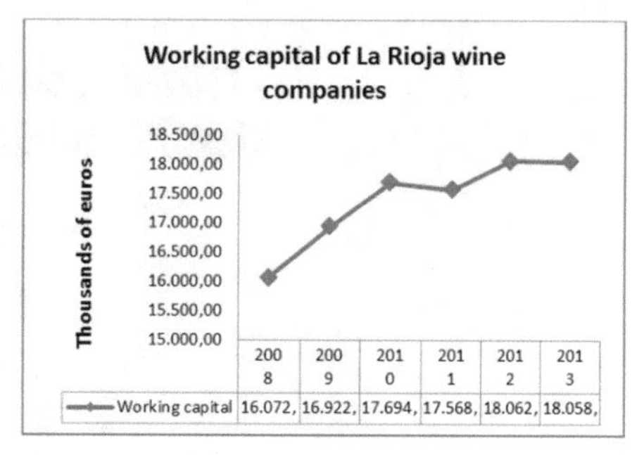

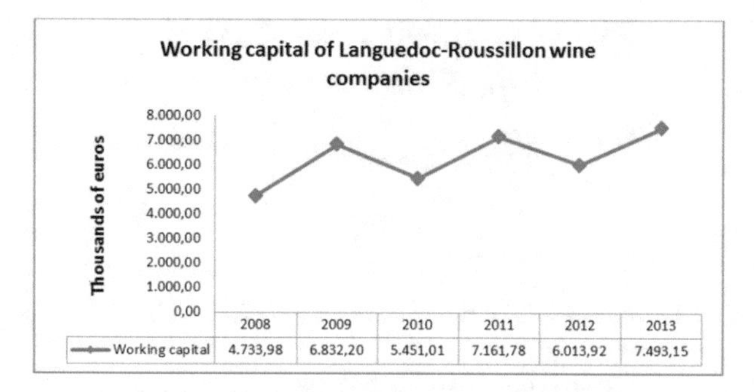

Fig. 2. Working capital of the three zones analysed in the period 2008–2013

4.2 Comparative of the Analysis of the Long-Term Financial Situation

The analysis of the long-term financial situation has as its main objective the measurement of the capacity that the company has to satisfy the long-term debts.

According to the criterion provided by Amat and Perramon (2012), in this period it is not possible to speak of a steady growth by the Catalan wine companies since there is no efficient management of the assets (the decrease of the sales is greater than the assets), nor a good management of the expenses (the results have dropped more than the sales), although it is necessary to emphasize a prudent financial management (the decrease of debts is greater than the assets) (Table 1).

On the other hand, in the case of La Rioja wine companies, in this period, you cannot speak of a steady growth: there is an efficient assets management (the decline in sales is lower than the assets) and a prudent financial management (the decline in debts is greater than the decline in assets); however, it is necessary to highlight that there is not a good expenses management (decrease in results is greater than in sales) (Table 2).

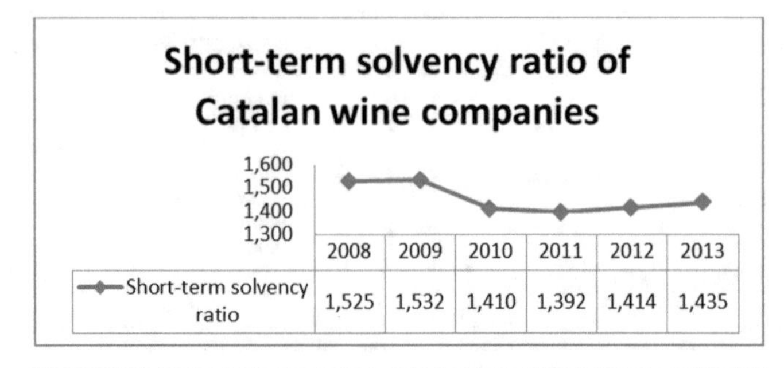

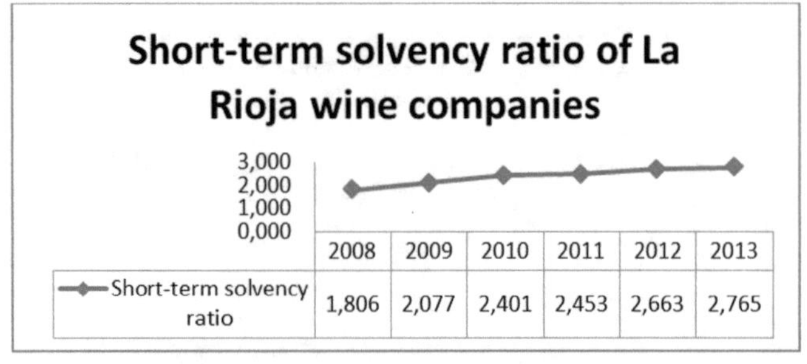

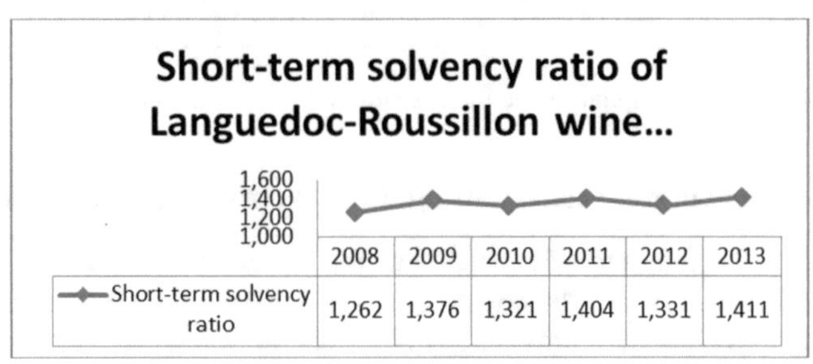

Fig. 3. Solvency at short-term of the three zones analysed in the period 2008–2013

Table 1. Balanced growth of Catalan companies in the period 2008–2013

	Asset management		Financial management		Expense management	
	Δ Sales	Δ Asset	Δ Asset	Δ Expense	Δ Expense	Δ Sales
Increasing of catalan companies	−14,59%	−6,05%	−6,05%	−8,25%	−60,17%	−14,59%

Table 2. Balanced growth of the companies of La Rioja in the period 2008–2013

	Asset management		Financial management		Expense management	
	Δ Sales	Δ Asset	Δ Asset	Δ Expense	Δ Expense	Δ Sales
Increasing of Rioja companies	−12,56%	−14,86%	−14,86%	−41,24%	−45,80%	−12,56%

Lastly, in the case of the Languedoc-Roussillon wine companies in this period, speaking of a balanced growth is permitted: there is an efficient management of the assets (the sales growth is greater than the assets), a good expenses management (the results have increased more than the sales) and, also a prudent financial management (debt growth is lower than the assets) (Table 3).

Table 3. Growth balanced of the Languedoc-Roussillon companies in the period 2008-2013

	Asset management		Financial management		Expense management	
	Δ Sales	Δ Asset	Δ Asset	Δ Expense	Δ Expense	Δ Sales
Increasing Llenguadoc-Rosselló companies	48,23%	24,91%	24,91%	16,77%	230,77%	48,23%

The vertical and horizontal percentages of the different financial items in the analysed period show that the net worth has increased in the sample of the Languedoc-Roussillon and La Rioja wine companies (they have been capitalized), but the Catalan wine companies have suffered a slight decapitalization.

On the other hand, the debts of the Catalan wine companies decreased by 8.24%, but the assets decreased by 6% and sales fell by 14.59%. However, in the case of the Languedoc-Roussillon winegrowers the opposite occurs, as the debts increased by 17%, the assets by 25%, and sales by 48%. In the case of winegrowers in La Rioja, debts decreased by 41%, assets decreased by 15% and sales fell by 13%.

In the case of the Catalan wine companies in the same way as in the wine companies of La Rioja, the net result decreases with important falls especially between 2008 and 2011; while in the case of the sample of the Languedoc-Roussillon there is an increase by 230% of the net result, with a strong increase in the period 2008–2010.

Following with the analysis of the total turnover of the asset, which informs us of the average time to recover the value of the asset, indicates that it takes the Catalan wine companies and the winegrowers of La Rioja less than a year to recover their investment value; while the wine companies of the Languedoc-Roussillon, the time limit is around a year. It must be said that in the period analysed there are not many differences between the three samples.

As for indebtedness, in the case of the sample of the Catalan and La Rioja wine companies, it can be seen that these are companies with an acceptable level of indebtedness, between 40% and 55%, but of poor quality, as there are more short-term debts (65%–80%) than in the long term (20%–35%). In the contrary, the sample of the Languedoc-Roussillon shows that these are companies with a high level of indebtedness, between 70% and 75%, and of poor quality, because short-term debts represent 75–86% and evidently the long term 14–25%.

The indebtedness in the three samples has a similar evolution (both in the Catalan wine sector, La Rioja and Languedoc-Roussillon), although it is worth mentioning the notable decrease in indebtedness in La Rioja during this period and the high indebtedness of the Languedoc-Roussillon wineries which is maintained throughout all the analysed period.

Table 4. Evolution of the financial indicators in the three areas of analysis in the period 2008–2013

Cataluna	2008	2009	2010	2011	2012	2013
Total equity	53.260,84	55.264,84	52.008,57	48.000,81	49.131,11	51.232,14
Liabilities	54.471,62	57.503,61	51.854,55	48.042,37	46.081,22	49.978,73
Asset	107.732,46	112.768,46	103.863,12	96.043,18	95.212,33	101.210,87
Sales	73.316,12	65.037,00	66.221,48	62.966,09	61.460,85	62.619,19
Results	1.847,58	340,27	887,29	207,74	1.078,09	735,84
La Rioja	2008	2009	2010	2011	2012	2013
Total equity	26.758,34	28.398,60	29.917,31	31.088,65	30.780,84	30.519,53
Liabilities	29.277,60	27.394,12	22.677,31	19.544,99	16.742,54	17.188,26
Asset	56.035,94	55.792,71	52.594,62	50.633,63	47.523,38	47.707,78
Sales	29.770,18	25.599,49	25.037,35	25.243,74	25.662,33	26.030,49
Results	2.112,31	3.585,45	2.166,52	1.446,92	2.107,24	1.144,85
Llenguadoc R	2008	2009	2010	2011	2012	2013
Total equity	7.378,94	9.379,62	9.193,41	9.594,03	9.603,83	10.927,96
Liabilities	21.016,44	22.909,05	20.808,62	23.203,88	23.155,90	24.541,67
Asset	28.395,38	32.288,68	30.002,03	32.797,90	32.759,73	35.469,68
Sales	27.490,17	29.564,41	27.834,78	33.184,42	36.375,48	40.748,68
Results	–250,93	–26,61	197,47	79,51	16,19	328,15

As for the debt quality ratio, there are similar values both in the case of the Catalan, La Rioja and Languedoc-Roussillon wineries, that is to say, more short-term debt than in the long term, and in Catalan ones, an improvement in the quality of the debt is experienced.

Therefore, we can say that the large Catalan and La Rioja wine companies have a good long-term financial position in the period considered, but it would be advisable to correct the poor quality of the debt in these generalized crisis years, renegotiating the debt, whenever possible, from short to long term. As for Languedoc-Roussillon companies, they have too much indebtedness and of poor quality in the period analysed (Table 4).

5 Economic Analysis

The economic analysis aims to explain how the results of these companies have been produced, identifying the causes which have provoked the results variation in the period under study.

Firstly, the value added of these wine companies in the three territories will be compared, as a difference between operating revenue and operating expenses, and as value of the income generated by the winemaking activity, of wine production by Catalan, La Rioja and Languedoc-Roussillon wineries. As shown (Fig. 3), the evolution of value added is quite divergent among the three samples: in the case of Catalan wine companies the trend is a decrease in the value added over the period analysed, being the period 2008–2009 and 2012–2013 the most decreasing one, and as for La Rioja companies there is also a decrease in the years 2012 and 2013, in contrast, in the case of the Languedoc-Roussillon wine companies the trend is the increase in the value added, being the 2012 and 2013 years the fastest growing ones.

As for the evolution of staff productivity, as a quotient between value added and personnel costs, a similar situation as in the case of added value analysed above is generated; the percentage of productivity is higher in the Languedoc-Roussillon wine companies, and in the Catalan and La Rioja wineries, it drops at the end of the period analysed, especially from 2011 onwards for Catalan ones and from 2012 onwards for La Rioja.

In regard to the profitability analysis of the Catalan wine companies, as it is conceived in Fig. 4, the return on equity (ROE) decreases until 2011, but in 2012 it experiences a considerable increase which moderates in 2013; the return on assets (ROA) is reduce by 79% in the period indicated. On the other hand, the return on equity (ROE) surpasses the return on assets (ROA), except in 2011,1 corroborating that the indebtedness harms the companies in the fiscal year 2011, unlike the rest of the years studied.

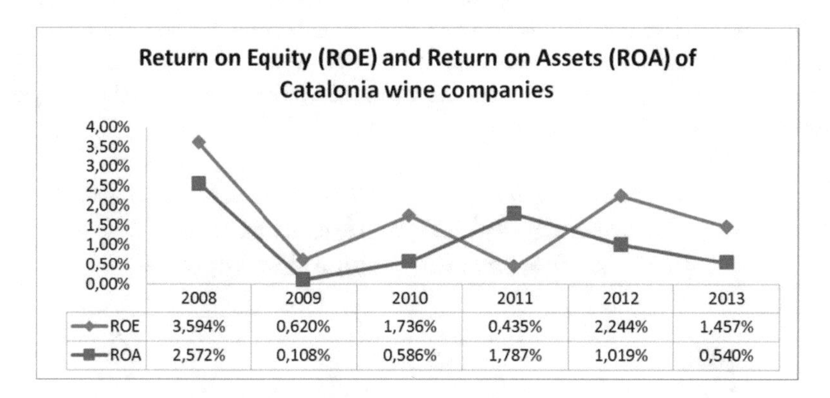

	2008	2009	2010	2011	2012	2013
ROE	3,594%	0,620%	1,736%	0,435%	2,244%	1,457%
ROA	2,572%	0,108%	0,586%	1,787%	1,019%	0,540%

Fig. 4. Financial and economic profitability of Catalan companies in the period 2008–2013

Lastly, in respect to the winegrowers of La Rioja, the profitability analysis, as shown in Fig. 5, indicates that the return equity (ROE) increased by 216% in 2009, but then from 2010 to 2013, it experiences a drop of 78%. However, the return on assets (ROA) remains more stable than the ROE and increases by 18% in the period 2009–2013. In the period 2008–2011, the ROE surpasses the ROA corroborating that the indebtedness does not harm the companies, although, in the years 2011 and 2012, it is below, indicating that in these years, the indebtedness does harm the company.

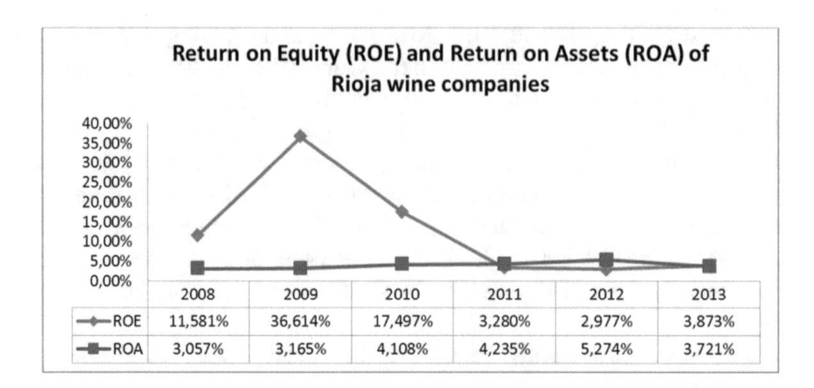

	2008	2009	2010	2011	2012	2013
ROE	11,581%	36,614%	17,497%	3,280%	2,977%	3,873%
ROA	3,057%	3,165%	4,108%	4,235%	5,274%	3,721%

Fig. 5. Financial and economic profitability of La Rioja ones in the period 2008–2013

In contrast, in the case of the analysis of the profitability of Languedoc-Roussillon wine companies (Fig. 6), apart from seeing that the financial profitability has lower values than Catalan wineries, a different evolution is also observed, as the return on equity (ROE) increases until 2010, but in 2011 it experiences a drop that later, in the year 2013, it recovers itself; the return on assets (ROA), however, remains more or less stable in the indicated period. But except for fiscal years 2008 and 2012, the ROE surpasses the ROA corroborating that the indebtedness does not hurt the companies, except for the years 2008 and 2012.

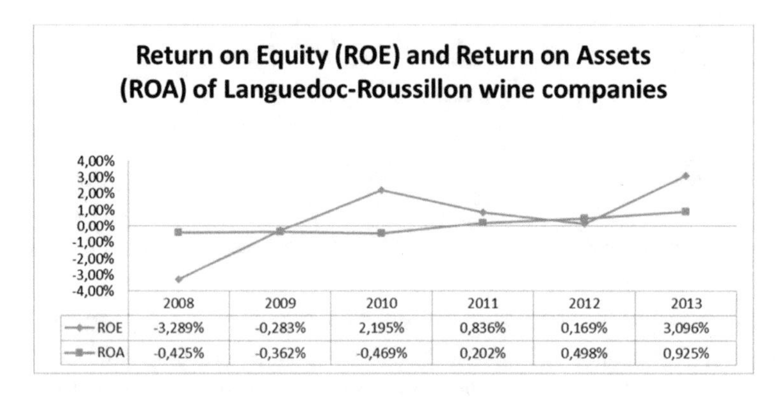

	2008	2009	2010	2011	2012	2013
ROE	-3,289%	-0,283%	2,195%	0,836%	0,169%	3,096%
ROA	-0,425%	-0,362%	-0,469%	0,202%	0,498%	0,925%

Fig. 6. Financial and economic profitability of the Languedoc-Roussillon in the period 2008–2013

Then, in regards to the net result, the behaviour in the three zones is totally different. While in the case of the Catalan and La Rioja wineries, it is more unstable with a tendency to decrease, in the case of Languedoc-Roussillon wineries the results remain more stable with a clear tendency towards growth.

It should be borne in mind that the highest median net result is from La Rioja wine companies although it is decreasing in the analysed period.

Finally, the evolution of net worth in the three areas must be analysed. In the case of Catalan wine companies, the net worth is high in comparison to the other areas analysed although the evolution of the net worth in the analysed period declines slightly. As for the La Rioja companies, the evolution of net worth in this period increases by 14%, that is, the companies of La Rioja are capitalized. With respect to Languedoc-Roussillon companies, there is a considerable increase in net worth (48%), which is why these companies are capitalized although they show a much lower equity/net worth than the Catalan and La Rioja companies.

6 Comparison of the Catalan, La Rioja and Languedoc-Roussillon Wine Companies in the Period 2008–2013

With the same criteria of sample selection for all the companies of the study, the large Catalan companies, which elaborate wine as main activity, show more investments in fixed assets or non-current assets than the others, that is to say, their economic structure in the analysed period 2008–2013 is divided equally between current and non-current assets, unlike the economic structure of the Rioja and Languedoc-Roussillon companies which show much lower non-current assets. Regarding the financial structure, Spanish companies are funded almost equally between self-financing and third-party funding; In contrast, French companies show more external funding, especially in the short term.

Comparing the analysis in the short term, it is emphasized that the working capital is positive in all companies, but it is especially notable for the companies of La Rioja, which have a high short-term solvency, so they can cope with short-term payments. Catalan and French companies show a more moderate short-term solvency, with no potential problems to meet short-term payment obligations.

The comparison of the long-term analysis indicates that Catalan companies should improve the management of assets and expenses, although they show a prudent financial management; with respect to the La Rioja companies, they do not show a balanced growth either, since, in spite of the assets and financial good management, they need to improve the management of the expenses; on the other hand, French companies have a balanced growth as they manage assets and expenses well, and have a prudent financial management. In regard to the recovery of the asset, Spanish companies recover the assets sooner than the French ones, and in terms of indebtedness, French companies are the most indebted (70% is indebtedness from third parties), but both Spanish and French companies should improve in quality of debt.

In regard to the economic analysis, which aims to explain the results of this period, it should be emphasized that the value added from the fiscal year 2011 grows for the French and decreases for the Catalan ones (due to the significant operating expenses derived from the investments) and for the case of the La Rioja companies, it decreases from the 2012. The results of the Spanish companies surpass the French ones, although it is necessary to continue valuing the change of trend which influences the fiscal year 2013, where the results of the Spanish companies decrease and of the French ones grows.

Regarding the analysis of profitability, La Rioja companies have the highest return on equity, although it declines considerably from 2011. In the case of Catalan companies the return on equity decreases until 2011 and grows afterwards, and for French companies, the return on equity exceeds that of the Catalan ones but it is inferior to those of La Rioja. The highest economic profitability is that of La Rioja companies followed by Catalan companies, but in 2013 there is a turning point which needs to continue being analysed.

Finally, it should be pointed out that the large wine companies in these areas are mostly family-owned companies, with high assets comparing the Spanish companies to French companies, with a large capitalization in La Rioja and Languedoc-Roussillon companies.

7 Conclusions

The large wine companies in Catalonia, La Rioja and Languedoc-Roussillon, in the crisis period analysed 2008–2013, show different results.

Catalan companies have more investments in non-current assets compared to the other areas analysed, and are the most capitalized companies, but they need to improve the assets and expenses management. Their indebtedness is acceptable, since it is at approximately 50%, and although they have decreased by 14.6% in sales, in this period, the results of the Catalan and La Rioja companies surpass the ones of the French, although it is still necessary to keep valuing the change in trend at the end of the period analysed.

La Rioja companies are the most solvent in the short term, they are companies capitalized with an acceptable indebtedness, and, they fell by 12.6% in sales in this period, but are the companies with the highest profitability, both the return on equity (ROE) and return on assets (ROA). The analysis shows that there is a need to improve the expenses management.

The Catalan and La Rioja companies in this period have innovated and have continued on the path of the exporting tradition opening new markets.

On the other hand, the large wine companies of Languedoc-Roussillon in this period have opted for investments in non-current assets, to respond to the innovation of the wine sector, with third-party funds and, although they have a good assets and expenses management show too high indebtedness, although they increase sales and results in this period. Thus, Languedoc-Roussillon has made significant investments with high short-term indebtedness, since it focuses on strategies of specialization and differentiation rather than low-cost strategies (Dukesneois et al. 2010).

The crisis period analysed is a period of changes in the financial economic health of wine companies due to the effects of innovation, internationalization and crisis in the three zones. In general, being large companies, in this period, the exporting tradition favours its economic financial health, and, despite being mostly family businesses, they are companies with differentials in products and processes.

References

Bardají, I., Estavillo, J., Iráizoz, B.: Export performance: insights on the Spanish wine Industry. Eur. J. Int. Manag. **8**(4), 392–414 (2014)

Barmo, M.: Enterprises and competitiveness: an Italian case study. Qual. Mang. **15**(142), 94–109 (2014)

Bresciani, S., Giacosa, E., Broccardo, L., Culasso, F.: The family variable in the French and Italian wine sector. EuroMed J. Bus. **11**(1), 101–111 (2016)

Broccardo, L., Giacosa, E., Ferraris, A.: The family variable in the wine sector: an Italian perspective. Int. J. Manag. Pract. **8**(3), 199–214 (2015)

Carneiro, A., Fensterseifer, J.E., Prévot, F.: The impact of export performance resources of companies belonging to clusters: a study in the French winery industry. Rev. Bus. Manag. **16** (52), 374–391 (2014)

Castillo-Valero, J.S., Garcia-Cortijo, M.C.: Análisis de los factores explicatives de la rentabilidad de las empreses vinícoles de Castilla-La Mancha. Revista FCA UNCUYO **45**(2), 141–154 (2013)

Dukesneois, F., Roy, F., Gurau, C.: Stratégies concurrentielles dans une industrie en crise Le cas de l'industrie vitivinicole en Languedoc-Roussillon. Revue française de gestion **203**, 41–56 (2010)

Francioni, B., Vissac, T., Musso, F.: Small Italian wine producers internationalization: the role of network relationships in the emergence of late starters. Int. Bus. Rev. (2016). http://dx.doi.org/10.1016/j.ibusrev.2016.05.003

Galbrealh, J.: A study of women in top business roles: the case of the wine industry. Int. J. Wine Bus. Res. **27**(2), 143–158 (2015)

Garcia-Galan, M.M., Moral-Agundez, A., Galera-Casquet, C.: Valuation and importance of the extrinsic attributes of the product from the firms perspective in a Spanish wine protected designation of origin. Span. J. Agric. Res. **12**(3), 568–579 (2014)

Giuliani, E.: Clusters, networks and firms product succés: an empirical study. Manag. Decis. **51** (6), 1135–1160 (2013)

Gómez-Limón, J.A., Casquet, E., Atance, I.: Análisis económico-financiero de las cooperativas agrarias en Castilla y León. Cirec España **46**, 151–189 (2003)

Marchini, A., Riganelli, Ch., Diotallevi, F., Paffarini, Ch.: Factors of collective reputation of the Italian PDO wines: an analysis on central Italy. Wine Econ. Policy **3**(1), 127–137 (2015)

Recasens, X., Alfranca, O., Maldonado, R.: The adaptation of urban farms to cities: the case of the Alella wine region within the Barcelona Metropolitan Region. Land Use Policy **56**, 158–168 (2016)

Roman-Sánchez, I., Aznar-Sánchez, J.A., Belmonte-Ureña, L.J.: Heterogeneity of the environmental regulation of industrial wastewater: European wineries. Water Sci. Technol. **72**(9), 67–72 (2015)

Sellers-Rubio, R., Alampi, V., Menghini, S.: Productivity growth in the winery sector: evidence from Italy and Spain. Int. J. Wine Bus. Res. **28**(1), 59–73 (2016)

Thomas, L.C., Painbeni, S., Barton, H.: Entrepreneurial marketing within the French wine industry. Int. J. Entrepreneurial Behav. Res. **19**(2), 238–260 (2013)

Vrontis, D., Bresciani, S., Giacosa, E.: Tradition and innovation in Italian wine family businesses. Br. Food J. **118**(8), 4–24 (2016). https://doi.org/10.1108/BFJ-05-2016-0192

FRvarPSO: A Method for Obtaining Fuzzy Classification Rules Using Optimization Techniques

Patricia Jimbo Santana[1], Laura Lanzarini[2], and Aurelio F. Bariviera[3(✉)]

[1] Facultad de Ciencias Administrativas, Carrera de Contabilidad y Auditoría, Universidad Central del Ecuador, Quito, Ecuador
prjimbo@uce.edu.ec
[2] Facultad de Informática, Instituto de Investigación en Informática, LIDI (Centro CICPBA), Universidad Nacional de La Plata, La Plata, Argentina
laural@lidi.info.unlp.edu.ar
[3] Department of Business, Universitat Rovira i Virgili, Avenida de la Universitat, 1, Reus, Spain
aurelio.fernandez@urv.cat

Abstract. FRvarPSO is a new method for obtaining classification rules, which operates on nominal and/or fuzzy attributes. It combines LVQ, which is a supervised learning neural network. The search is performed through an optimization technique such as varPSO, considered a metaheuristic based on particles clusters of variable population. Each individual represents a possible solution to the problem. The proposed method uses a voting criterion, which affects the particle's speed. This method is benchmarked against PART and C4.5, on 12 databases of the UCI repository and three databases of financial institutions from Ecuador. The results obtained were satisfactory.

Keywords: Fuzzy Rules · varPSO · Particle Swarm Optimization · Classification rules · Data mining

1 Introduction

Data mining is considered as the process of extracting knowledge from large databases. Knowledge generated could be of great interest to organizations, since it could allow them to make informed decisions, generating a competitive advantage.

Among the set of techniques present in data mining, we can mention fuzzy logic. This technique is used to model imprecise data, and allows the handling of uncertainty. Classification rules constitute a set of techniques that help the decision making process. Classification rules using fuzzy sets to describe the values of their attributes, are called fuzzy classification rules. The method proposed in this article is called Fuzzy Rules

J. C. Ferrer-Comalat et al. (Eds.): MS-18 2018, AISC 894, pp. 112–126, 2020.
https://doi.org/10.1007/978-3-030-15413-4_9

variable Particle Swarm Optimization (FRvarPSO), which allows the generation of fuzzy classification rules. A fuzzy classification rule is an expression, whose statement is given as follows:

$$\text{IF } condition1 \text{ } and \text{ } condition2 \text{ THEN } condition3$$

where *condition1* and *condition2* are attributes of the antecedent of the rule. They can include nominal and/or fuzzy variables, of the form (attribute = value). The attribute *condition3*, present in the consequent of the rule, represents a single attribute, which can assume different values, each one of them representing a class. The only restriction that these rules present is that attributes that intervene in the antecedent of the rule should not be not part of the consequent (Ganeshkumar et al. 2013). Fuzzy variables use linguistic terms that are characterized by a fuzzy set. It is important to indicate that the smaller the number of rules to analyze, the easier to interpret the model will be.

The contribution of this article is the proposal of a method of obtaining fuzzy classification rules, combining a competitive neural network, with a variable population optimization technique. Additionally, a voting criterion on the membership degree of the attributes to the fuzzy set is incorporated. Its goal is to obtain a model that combines accuracy, with a reduced number of classification rules. We would like to highlight that a reduced rule set facilitate model interpretation.

This article is organized as follows: Sect. 2 outlines related works, Sect. 3 describes the proposed method, Sect. 4 shows the results obtained. Finally, Sect. 5 concludes and proposes some future research lines.

2 Related Works

Data mining uses several methods for knowledge extraction, such as artificial neural networks, fuzzy set theory, decision trees, support vector machines, and genetic algorithms, among others (Witten et al. 2016).

Regarding neural networks, there are different architectures depending on the type of problem to solve. These architectures include popular models, such as backward propagation networks, Self-Organizing Maps (SOM) and Learning Vector Quantification (LVQ), which are neural networks that use supervised learning (Hasperué et al. 2005).

In the case of classification rules, solutions can be found structured in the form of trees, as is the case of method C4.5 (Quinlan 1993), which generates a classification tree, whose branches allow obtaining the set of rules, another method is PART (Eibe and Witten 1998). In the previous cases, the important thing is to obtain a set of rules that cover the examples, considering a pre-established error level.

There is also the case of Support Vector Machines. In this method, depending on the type of discriminant function that is used; it is possible to build extremely powerful linear and nonlinear models.

Another set of methods uses elements from the theory of fuzzy sets, developed from the seminal work by Zadeh (1965). This technique is very useful, mainly in situations where the boundaries of the variables under analysis are not well defined. In such cases, the use of fuzzy attributes, which are formed by linguistic terms, is necessary.

The literature on classification methods covers also other types of techniques, such as genetic algorithms (GA), ant colonies (ACO), and optimization based on particle clusters (PSO). These are population optimization techniques, inspired by different biological processes (Ziqiang et al. 2007; Venturini 1993).

In previous research, we have developed a model that combines a competitive neural network, with supervised learning LVQ (Learning Vector Quantization) based on centroids or prototypes, with a technique of optimization by particle cluster of variable population varPSO (Lanzarini et al. 2015a, b, Lanzarini et al. 2017).

The novelty of this work is the introduction of fuzzy variables, used to improve the solution through a variable population optimization technique, which is used in the movement of the particle, considering a voting criterion. This change allows the generation a smaller number of rules, while improving the classification accuracy. Fuzzy classification rules obtained are formed by nominal and/or linguistic variables.

3 Methodology

This paper presents a hybrid methodology, which combines a competitive neural network with an optimization technique based on particles' clusters. We aim to obtain a set of classification rules, whose antecedent is formed by nominal and/or fuzzy variables. Before describing the proposed method, we briefly describe some concepts upon which our methodology is based.

3.1 Fuzzy Variables

Fuzzy Logic was introduced by Zadeh (1965). It allows to model the uncertainty that exists in real-world problems. It also handles imprecise information in a similar way as in human reasoning. Knowledge can be expressed linguistically, with different degree of precision. Fuzzy Logic is based on the fact that a statement must not only be true or false, but it can be true up to a certain degree. It uses membership functions, in order to indicate the degree of belonging of the variable to a given fuzzy set. An excellent introduction to fuzzy logic and applications can be consulted in Yager and Zadeh (2012).

Fuzzy Sets can be used to describe linguistic terms such as "high", "medium", and "low". The Fuzzy Set is characterized by the variable, which has a membership function to that set, which is known as the degree of membership having a real value within the interval [0–1]. If the statement is completely true, it will have the value of 1; and if it completely false, it will have a value of 0. For example, let consider the case of imprecise data in the credit analysis. The numerical variable "Income = USD 4.000.00", can be considered as "High income with a membership degree of 0.3", "Average income with a membership degree of 0.6", and "Low income with a membership degree of 0.1". Consequently, it indicates that this income value is not fairly considered as low income.

Fuzzy Inference Systems (FIS) use fuzzy sets theory, in order to make the correspondence between the linguistic variables (input), and the nominal variable (output).

Fuzzy classification rules are rules that have linguistic sentences that describe how the FIS makes decisions about a set of input variables. These classification rules have an antecedent formed by nominal and/or fuzzy variables, and a consequent (knew in advance) called "class". When the antecedent of the rule is formed by variables that use the conjunction operator for several conditions, the minimum operator or product can be used between the degrees of membership of the variables.

Fuzzy logic has allowed researchers to obtain ways to improve the performance of metaheuristics, to accelerate convergence and/or obtain better quality solutions.

3.2 Learning Vector Quantization (LVQ)

In this work, two types of competitive neural networks are used: SOM (Self Organization Map) and Learning Vector Quantization (LVQ). These two types of networks have the objective of grouping the input data and presenting, as a result, a set of centroids. It is important to keep in mind that each of the centroids tries to represent a set of input data, considering a measure of distance that was previously defined.

In addition to the clustering of available information, SOM is characterized by its ability to preserve input data (Kohonen 2012); it can be represented as a two-layer structure: the entry layer, which has the function of allowing the entry of information to the network, and a competitive layer that, is in charge of carrying out the grouping. Each competitive neuron is associated to vector of weights, represented by the values of the arcs that reach it through the input layer. In this way, SOM manages two information structures: the first one has information of centroids associated with competitive neurons and the other one is responsible for determining the proximity between neurons. SOM is in charge of providing additional information regarding the way the groupings are organized.

LVQ is a supervised classification algorithm. Different variants of LVQ can be consulted in Aggarwal (2015). It is characterized because each neuron of the competitive layer has a numerical vector of the same dimension of the input examples and a label that indicates its class. These vectors are the ones that, at the end of the adaptive process, contain the information of the centroids or prototypes of the classification. When the algorithm starts, a number K of centroids to be used is indicated. This allows to define the architecture of the network, where the number of inputs and outputs are defined according to the problem to solve. The centroids are initialized taking K random examples. Then, examples enter and adopt centroids' position. Subsequently, the closest example centroid is determined, using a preset distance measure. Since it is a supervised process, it is possible to determine whether the example and the centroid correspond to the same class or not. If the centroid and the example belong to the same class, the centroid "approaches" the example with the aim of strengthening its representation. Otherwise, the centroid "moves away".

This process is repeated until the changes to be performed are under (or above) a threshold, or until the examples are identified with the same centroids in two consecutive iterations, whichever first occurs. In addition, the second closest centroid is also considered, provided that the class to which it belongs is different from the default class, and is less than 1.2 times the distance from the first one, due to the inertial factor that was established previously. LVQ, when using information of the expected response during training, presents a better fit of the centroids. In this work, we use both architectures, being LVQ the one that presented better performance in the development of the method.

3.3 Particle Swarm Optimization (PSO)

Particle Swarm Optimization (PSO) is a population metaheuristic, proposed by Kennedy and Eberhart (1995), where each individual of the population, called a particle, represents a possible solution to the problem and adapts following three factors: (i) its "aptitude", which is the knowledge about the environment; (ii) its "memory", which is given by the historical knowledge or previous experiences; and (iii) the "social" knowledge, which is the historical knowledge or previous experiences of the individuals located in their neighborhood. There are different PSO variants that control the way in which particles move within the solutions space (Ganeshkumar et al. 2013; Chen 2006).

In this work, PSO was used in order to obtain fuzzy classification rules, able of operating on nominal and/or fuzzy attributes. This is done by combining the previously mentioned methods is required, indicating that nominal and fuzzy attributes will be part of the background of the rules. In the case of fuzzy attributes, it is necessary to know the membership degree to each of the fuzzy sets. Additionally, the Iterative Rule Learning (IRL) approach is followed (Venturini 1993). According to IRL, each individual represents a single rule. The solution to the problem is constructed from the best individuals, obtained in a sequence of executions. This population technique is applied in an iterative way, until achieving the desired coverage. At the end of each iteration, a single fuzzy classification rule is obtained, represented by the best individual of the population.

We use a fixed-length representation, in which only the antecedent of the rule is coded. Considering our approach, an iterative process will be carried out associating all the individuals of the population with a predetermined class, which does not require the coding of the consequent. The effectiveness of these population techniques are based on the size of the population. Therefore, a method with a variable population strategy is introduced. We start the process with a minimum population size. The, we adjust the number of particles during the adaptive process. PSO uses the nominal or fuzzy variables, reducing the number of attributes of the antecedent. The consequent of the rule is given by the resulting class. In addition, we introduce a "voting" criterion. Each time the fitness function is evaluated, the membership degree of the examples that meet the rule is averaged. This information is used to optimize the movement of the individual particle. This movement is about the election in global terms, because the membership degrees of the examples that meet the rule are considered, giving priority to movement in the direction towards the highest value is found.

3.4 Proposed Method FRvarPSO for Obtaining Fuzzy Rules

The proposed method uses PSO to generate classification rules that can operate on nominal and/or fuzzy attributes. It combines the methods indicated above, for the selection of the attributes that will form the antecedent and the range that these attributes should take.

Fuzzy Attributes

The method begins by transforming the numerical variables into linguistic variables with three possible terms: LOW, MEDIUM and HIGH. Each of these terms is determined by a fuzzy set and with a membership function. The fuzzy sets *A in X* are considered, associated with a membership function $U_A(x)$ where:

$$U_A(x): x \rightarrow [0,1]$$

The membership function $U_A(x)$, assigns a value between 0 and 1 to each x. The membership functions associated with each of the fuzzy sets are triangular:

$$U_{Triangular}(x;a,b,c) = \begin{cases} 0 & x \leq a \\ x - \frac{a}{b-a} & a \leq x \leq b \\ c - \frac{x}{c-b} & b \leq x \leq c \\ 0 & c \leq x \end{cases}$$

We consider uniform fuzzy partitions, with the same number of labels for each of the analyzed variables. Consequently, the upper and lower limits of each variable was determined, and then it was uniformly distributed in nine intervals. The first fuzzy set was formed by the first four intervals. The second fuzzy set goes from the middle of the third interval to the middle of the seventh interval. The third fuzzy set is formed from the sixth to the ninth interval. In this way, each of the fuzzy sets has a uniform distribution, and there is an overlap on the borders, representing what happens in human reasoning.

Each numerical variable have, at most, two degrees of membership in two of the fuzzy sets.

To be able to operate with qualitative and/or fuzzy attributes, the movement of the particle is divided into a binary part and a real part. The binary part indicates the attributes that are going to be used. The real part, indicates the intervals of the fuzzy attributes.

Qualitative attributes are represented in binary form, because neural networks work only with numerical attributes. These attributes are encoded in such a way that each of them has as many binary digits as different values have the variable.

In the real part, the qualitative variable will have the value of 0 and the numeric variable contains the values of either the upper or lower limit, with the value of the membership degree to the fuzzy set of corresponding variable.

Subsequently, the training examples are grouped through a competitive neural network. We use the Euclidian distance as measure of similarity. The supervised adaptive strategy is done by means of a LVQ or SOM networks. The objective of using one of these neural networks is to select the most promising areas within the search space. Such result will be used in PSO either to form the initial population, or when new individuals are generated to increase the population size.

Representation of Rules

We use the IRL approach in order to represent the rules. It uses a iterative process, in such a way that for each one of the executions the best individual is extracted. To avoid premature convergence to a single individual, the examples that were correctly covered, are removed from the set of entry examples, once the rule has been obtained.

Representation of the Antecedent and Consequent of the Rule

Regarding the representation of the antecedent, each one of the conditions is an attribute. The attributes can be nominal or fuzzy. In order to operate with these types of attributes, we use a fixed length representation. Such representation determine which attributes make up the rule, and their value or range of values. This representation is formed by two parts.

- The Binary Part will use, for the nominal attributes, sets of binary digits for the different values of the attribute. Alternatively, for each of the fuzzy attributes, it will use only two binary digits.
- The Real Part contains the values that allow delimiting the fuzzy attributes when forming the condition. These values are calculated through the optimization technique. They have the same length as the binary part, used specifically for the fuzzy attributes.

To start executing the algorithm, the majority class is chosen. This class is preserved during the iterative process. Consequently, it will be not necessary to codify the consequent of the rule in the individual codification.

The network's size is variable depending on the type of PSO that is used. For the initialization of the population, the particles are created depending on the number of neurons; and are initialized using the previously trained neural network. The examples that have not been covered by each of the classes are analyzed. The process starts with the class that with the largest number of elements. Then there is an iterative process using PSO.

Structure of a Particle

To move in an *n*-dimensional space, each particle *p* is formed by:

- $pBin_i = (pBin_{i1}, pBin_{i2}, pBin_{i3}, \ldots, pBin_{in})$ stores the current position of the particle. In our case, are the conditions that form the rule's antecedent.
- *Vectors Velocity* $V1_i = (V1_{i1}, V1_{i2}, \ldots, V1_{in})$ y $V2_i = (V2_{i1}, V2_{i2}, \ldots, V2_{in})$. They combine to indicate the direction towards the particle will move. This is a real vector.
- Best Position Vector ($pBestBinario_i = (pBestBin_{i1}, pBestBin_{i2}, pBestBin_{i3}, \ldots, pBestBin_{in})$ $p\text{BestBin}_{\text{i1}}$, $p\text{BestBin}_{\text{i2}}$, $p\text{BestBin}_{\text{i3}}, \ldots,$ $p\text{BestBin}_{\text{in}}$) stores the best solution found by the particle so far. This is a binary vector.
- *Fitness* X_i stores the fitness value of the current solution. Vector X_i.
- *Fitness* $pBest_i$ stores the fitness value of the best local solution found so far. Vector $pBestBin_i$.
- $pReal_i = (pReal_{i1}, pReal_{i2}, pReal_{i3}, \ldots, pRealn_{in})$ is used for fuzzy attributes containing the current bounds of the intervals, indicating the membership degree to the fuzzy set.

- Real velocity $VelocReal_i$ $V3_i = (V3_{i1}, V3_{i2}, ..., V3_{in})$ Velo indicates the direction of the change of the individual.
- $pBestReal_i = (pBestReal_{i1}, pBestReal_{i2}, pBestReal_{i3}, ..., pBestReal_{in})$ stores the best solution found by the particle within the limits of the intervals.
- $sopBin_i = (sopBin_1, sopBin_2, sopBin_3, ..., sopBin_n)$ stores the conditions that make up the antecedent of the rule, which represents the particle.

To initialize the population, positions and velocities are randomly generated, once the population has been created considering the centroids of the neural network. When the *i*-th particle moves, the current position is modified and the intervals corresponding to the numerical attributes change as follows:

Binary Part

We use $pBin_{i1}$ in order to identify the attributes forming the antecedent of the binary rule vector. Its movement is controlled through two velocity vectors $V1_i$ *y* $V2_{\bar{i}}$. $pBestBin_i$ and the closest solution within the search space are also taken into account in this way:

$$V1_{ij}(t+1) = W_{bin}V_{ij}(t) + \varphi 1\, rand_1 \cdot \left(2pBestBin_{ij} - 1\right) + \varphi 2\, rand_2 \cdot \left(2lBestBin_{ij} - 1\right)$$

- W_{bin} is the inertia factor, which decreases along the iterations of the algorithm.
- $\varphi 1, \varphi 2$ are constant values that represent the cognitive and social factors.
- $rand_1, rand_2$ are randomly defined values in the interval [0, 1].
- $pBestBin_{ij}$, $lBestBin_{ij}$ are the j-*th* vectors' digit
- $pBestBin_i$, $lBestBin_i$. Each particle takes into account the position of its closest neighbor, whose fitness value is superior to its own. Therefore, the value $lBestBin_i$, corresponds to the vector of $pBestBin_j$ of the closest particle to $pBestBin_i$, provided that it meets $Fitness_j > Fitness_i$ (measured as the Euclidean distance).

The values $W_{bin}, \varphi 1, \varphi 2$ are those that lead to the convergence algorithm. Each element of the velocity factor $V1_i$ is controlled in the following way:

$$V1_{ij}(t) = \begin{cases} \delta 1_j & \text{if } V1_{ij}(t) > \delta 1_j \\ -\delta 1_j & \text{if } V1_{ij}(t) \leq \delta 1_j \\ V1_{ij}(t) & \text{otherwise} \end{cases}$$

where,

$$\delta 1_j = \frac{limit1_{superiorj} - limit1_{inferiorj}}{2}$$

The velocity vector $V1_i$ is calculated according to $V1_{ij}(t+1)$ and is controlled according to $V1_{ij}(t)$. The value of the velocity vector is updated ($V2_i$) according to:

$$V2_{ij}(t+1) = V2_{ij}(t) + V1_{ij}(t+1)$$

$V2_i$ is controlled considering $\delta 2_j$:

$$\delta 2_j = \frac{limit2_{superiorj} - limit2_{inferiorj}}{2}$$

The new position of the particle is used as an argument of the sigmoid function given by:

$$sig(X) = \frac{1}{1 + e^{-x}}$$

The vector position of the particle is updated:

$$pBin_{ij}(t+1) = \begin{cases} 1 & if\ rand_{ij} < sig(V2_{ij}(t+1) \\ 0 & otherwise \end{cases}$$

Where $rand_{ij}$ is a random number, following a uniform distribution between [0, 1], and $sig(V2_{ij})$ is the probability that the element X_{ij} belonging to the vector that contains the current position becomes 1.

Continuous Part

This part controls the limits of the fuzzy attributes that intervene in the rule. It is calculated by adding to $PReal_i$ (vector position) the value of the velocity vector $V3_i(t)$. It is necessary to verify that the allowed limits are not exceeded.

$$pReal_{ij}(t+1) = pReal_{ij}(t) + V3_{ij}(t+1)$$

$$\begin{aligned} V3_{ij}(t+1) = & W_{real} V3_{ij}(t) + \varphi 3\, rand_3 \cdot \left(pBestReal_{ij} - pReal_{ij}\right) \\ & + \varphi 4\, rand_4 \cdot \left(lBestReal_{ij} - pReal_{ij}\right) \end{aligned}$$

- W_{real} is the inertia facto.
- $\varphi 3, \varphi 4$ are constant values, which represent the cognitive and social factors.
- $rand_3$ *y* $rand_4$ are random values within the interval [0, 1].
- $pBestReal_{ij}$, $lBestReal_{ij}$ are the j-th of the vectors $pReal_j$ and $lBestReal_j$, It is important to consider that each particle takes into account the position of its closest neighbor whose fitness value is higher than its own, so the value of $lBestReal_i$, corresponds to $pReal_j$ of the particle that is closest to $pReal_i$ provided $Fitness_j > Fitness_i$ (measured as the Euclidean distance).

It is important to indicate that in continuous PSO, at the beginning, the speed takes high values in order to explore the space of solutions. Then its speed decreases, so that the particle stabilizes. In binary PSO the inverse process happens. The extreme values mapped by the sigmoid function produces similar probability values, close between 0 and 1, reducing the probability of change in the values of the particle. If the velocity values are close to 0, then the probability of change is increased. If the velocity of a particle is determined by a null vector, then each of the binary digits that determine its position has a probability equal to 0.5 of changing to 1.

In this work, we considered the vector speed $V1$ limited in the range [–0.5, 0.5]. In the case of the variable $V3$, it was stablished between [–3, 3]. The values of $\varphi 1, \varphi 2, \varphi 3, \varphi 4$ for the binary and continuous parts, take the value 0.25. In the case of the inertia factors, W_{bin}, W_{real}, they were established between 1.25 and 0.25 in descending order for W_{bin} and in ascending order for W_{real}.

The inertia used to update the velocity vectors is calculated as follows:

$$W = W_{final} + \left(W_{initial} - W_{final}\right) * \frac{Current\ Iteration}{Total\ Iterations\ Number}$$

The fitness function is updated according to the new position of the particle. If this fitness value is the best until that iteration, the best position vector $pBest_i$ is updated. The fitness function for each particle is calculated as follows:

$$Fitness\ Function = ((support/ExamplesNumber) * confidence * factor1) - (factor2 * AtribsAntecedentNum/MaxAtribs)$$

support and *confidence* are the values corresponding to the rule that represents the particle; *factor1* is a penalty value in the case that the support is not within the established ranges in the algorithm. The second term in the fitness function reflects the importance given to the number of attributes that form the antecedent, being *factor2* a constant.

Once the first rule has been obtained, the membership degree of each one of the examples meeting the rule Gp_i is obtained. Gp_i is the membership degree of example i that meets the rule. It is the result of a t-norm, that uses the minimum operator, between the memberships degrees of the fuzzy attributes involved in the antecedent of the rule.

Subsequently, the Voting Criteria (CV) is calculated. It is obtained as the average of the membership degrees of the examples that meet the rule. In other words, CV are determined by:

$$CV = \frac{\sum_{i=1}^{n} Gp_i}{n}$$

CV is used in the movement of the subsequent particle. Thus, the speed of j-th particle in the continuous part is updated in the following way:

$$\begin{aligned} V3_{ij}(t+1) = & W_{real} V3_{ij}(t) + \varphi 3\, rand_3 \cdot \left(pBestReal_{ij} - pReal_{ij}\right) \\ & + \varphi 4\, rand_4 \cdot \left(lBestReal_{ij} - pReal_{ij}\right) + CV_{ij} \end{aligned}$$

The voting criterion used in the selection of fuzzy attributes is taken into consideration for the individual's movement, taking into account that individuals will have (as a priority) the movement in the direction where the highest value is found, meaning the value that had the highest “vote”.

When a rule has been obtained, that set of examples that has been covered by such rule is removed from the input database. The process is iteratively performed until the maximum number of iterations is reached, or until it covers all the examples or until the number of examples of each of resulting classes are considered minimal, whichever comes first. When a new example is to be classified, the rules must be applied in the order in which they were obtained. Then, the example will be classified with the class corresponding to the consequent of the first rule whose antecedent is verified.

Figure 1 contains the pseudocode of the proposed method.

Determine the fuzzy sets corresponding to each of the values of the linguistic variables.
Determine the membership degree of the values of the numerical variables of all the training examples
Train the competitive neural network using the training examples
Determine the minimum support for each class
While *(The termination criterion is not reached)*
    Choose the class that has the largest number of examples that have not been covered
    Build a population considering the centroids of the neural network
    Evaluate the fitness value of each particle
    While the particle population does not stabilize
    Identify the best solution found so far.
    For each *particle*
        Calculate the voting criteria (average degree of membership of the examples that meet the rule).
        Calculate the speed and add the voting criteria mentioned previously.
        Obtain the new position of the particle by adding the previous speed and delimit accordingly.
        End for
    In case of using elitism, recover the best solution from the previous iteration.
    End while
    Obtain the best population rule
    If *the rule meets the support and confidence* ***then***
        Incorporate the rule into the rule set
        Remove correctly covered examples from the input set
        Recalculate the minimum support of the class that has been considered
    End if
End while

Fig. 1. Pseudocode of the proposed method

4 Results

In this section, we compare the performance of the proposed method FRvarPSO, with the methods C4.5 by Quinlan (1993), and PART (Eibe and Witten 1998). The indicated methods allow obtaining classification rules. C4.5 is a pruned tree method, whose branches are mutually exclusive and allow classifying the examples. PART method results in a list of classification rules equivalent to those generated by the proposed method but in a deterministic manner. The operation of PART is based on the construction of partial trees. Each tree is created in a similar way to the one proposed for C4.5. However, during the construction process errors are calculated for each branch. This feature allows the selection of the most convenient combinations of attributes.

The proposed method FRvarPSO obtains fuzzy classification rules, whose antecedent is formed by fuzzy and/or nominal variables, based on the combination of competitive neural networks (SOM and LVQ) and optimization techniques (PSO and varPSO).

Table 1. Classification accuracy obtained with different classification methods.

Data set	C4.5	PART	SOM + PSO	*SOM + Fuzzy PSO*	SOM + varPSO	*SOM + Fuzzy varPSO*	LVQ + PSO	*LVQ + Fuzzy PSO*	LVQ + varPSO	*FR varPSO*
Adult_data	0,8517 (0,0019)	0,8436 (0,0025)	0,8001 (0.0012)	*0,8461 (0.0046)*	0,7953 (0.0034)	*0,7977 (0,0660)*	0,8156 (0.0053)	*0,8239 (0,0900)*	0,8046 (0.0041)	*0,8118 (0,0047)*
Balance_scale	0,7516 (0,0120)	0,8105 (0,0019)	0,7317 (0,0090)	*0,7326 (0,0201)*	0,7222 (0,0045)	*0,7246 (0,0216)*	0,7421 (0,0011)	*0,7460 (0,0192)*	0,7325 (0,0011)	*0,7619 (0,0015)*
Breast_w	0,9556 (0,0076)	0,9525 (0,0081)	0,9454 (0,0101)	*0,9344 (0,0068)*	0,9526 (0,0067)	*0,9401 (0,0140)*	0,9457 (0,0088)	*0,9544 (0,0114)*	0,9512 (0,0072)	*0,9565 (0,0063)*
Credit_a	0,8524 (0,0036)	0,8494 (0,0027)	0,8507 (0,0186)	*0,8550 (0,0131)*	0,8658 (0,0080)	*0,8657 (0,0098)*	0,8630 (0,0077)	*0,8578 (0,0109)*	0,8635 (0,0088)	*0,8689 (0,0122)*
Credit_coop	0,8105 (0,0011)	0,8054 (0,0023)	0,7718 (0,0019)	*0,7857 (0,0026)*	0,7701 (0,0023)	*0,7891 (0,0021)*	0,7871 (0,0031)	*0,7922 (0,0029)*	0,7894 (0,0023)	*0,7988 (0,0129)*
Credit_g	0,7040 (0,0112)	0,7110 (0,0101)	0,6997 (0,0019)	*0,7536 (0,0106)*	0,6935 (0,0078)	*0,7697 (0,0081)*	0,6958 (0,0126)	*0,7578 (0,0091)*	0,6914 (0,0151)	*0,7592 (0,0058)*
Credit_bank1	0,9778 (0,0003)	0,9761 (0,0007)	0,9688 (0,0033)	*0,9819 (0,0037)*	0,9700 (0,0026)	*0,9840 (0,0032)*	0,9737 (0,0037)	*0,9869 (0,0024)*	0,9778 (0,0018)	*0,9880 (0,0026)*
Credit_bank2	0,8492 (0,0010)	0,8413 (0,0095)	0,8133 (0,0024)	*0,8316 (0,0022)*	0,8264 (0,0041)	*0,8385 (0,0033)*	0,8356 (0,0028)	*0,8455 (0,0029)*	0,8415 (0,0035)	*0,8501 (0,0033)*
Diabetes	0,7414 (0,0101)	0,7434 (0,0325)	0,7314 (0,0146)	*0,7325 (0,0136)*	0,7207 (0,0104)	*0,7269 (0,0158)*	0,7362 (0,0185)	*0,7369 (0,0164)*	0,7213 (0,0165)	*0,7442 (0,0148)*
Drugte	0,9300 (0,0301)	0,9075 (0,0112)	0,6300 (0,0010)	*0,8480 (0,0217)*	0,6150 (0,0106)	*0,8435 (0,0195)*	0,7950 (0,0035)	*0,8501 (0,0139)*	0,8409 (0,0336)	*0,8513 (0,0195)*
DrugY	0,9275 (0,0035)	0,9055 (0,0069)	0,8178 (0,0389)	*0,7818 (0,0342)*	0,7537 (0,0136)	*0,7846 (0,0275)*	0,7666 (0,0141)	*0,7899 (0,0267)*	0,8275 (0,0106)	*0,8405 (0,0247)*
Heart_c	0,7601 (0,0442)	0,7939 (0,0125)	0,7277 (0,0019)	*0,7698 (0,0198)*	0,7617 (0,0071)	*0,7663 (0,0266)*	0,7576 (0,0118)	*0,7983 (0,0024)*	0,7614 (0,0259)	*0,7866 (0,0248)*
Heart_Statlog	0,7666 (0,0274)	0,7333 (0,0276)	0,7497 (0,0028)	*0,7664 (0,0318)*	0,7772 (0,0145)	*0,7886 (0,0113)*	0,7611 (0,0131)	*0,7666 (0,0335)*	0,7957 (0,0219)	*0,7962 (0,0236)*
Wines	0,8763 (0,0102)	0,8820 (0,0126)	0,8749 (0,0029)	*0,8722 (0,0309)*	0,8833 (0,0118)	*0,8916 (0,0235)*	0,8778 (0,0079)	*0,8811 (0,0071)*	0,8833 (0,0157)	*0,8944 (0,0109)*
Zoo	0,9207 (0,0144)	0,9227 (0,0134)	0,9300 (0,0015)	*0,9326 (0,0275)*	0,9500 (0,0018)	*0,9567 (0,0025)*	0,9432 (0,0051)	*0,9588 (0,0102)*	0,9526 (0,0045)	*0,9601 (0,0023)*

Table 2. Average number of rules and standard deviation, obtained using different classification methods.

Data set	C4.5	PART	SOM + PSO	*SOM + Fuzzy PSO*	SOM + varPSO	*SOM + Fuzzy varPSO*	LVQ + PSO	*LVQ + Fuzzy PSO*	LVQ + varPSO	FRvarPSO
Adult_data	563,5000 (3,2510)	836,4232 (7,3026)	3,9420 (0.3536)	*2,7928* *(0.2828)*	3,2496 (0.0707)	*2,7919* *(0,2828)*	4,0987 (0.1414)	*2,7386* *(0,3536)*	4,0422 (0.3536)	*3,0106* *(0,2267)*
Balance_scale	52,0015 (1,2478)	47,0051 (1,3102)	7,8943 (0,4243)	*8,6188* *(0,4596)*	8,7994 (0,1414)	*8,2277* *(0,4451)*	8,3486 (0,2121)	*8,2038* *(0,3348)*	9,2498 (0,0707)	*8,1523* *(0,5794)*
Breast_w	11,0129 (0,2150)	11,1290 (0,1201)	2,7151 (0,1701)	*2,6480* *(0,1080)*	2,6430 (0,2014)	*2,5720* *(0,2150)*	2,7381 (0,1075)	*2,4977* *(0,1247)*	2,6763 (0,1476)	*2,3256* *(0,1567)*
Credit_a	17,1100 (0,1923)	37,0012 (0,1847)	3,0099 (0,0316)	*3,0083* *(0,0009)*	3,0099 (0,0118)	*3,0000* *(0,0015)*	3,0000 (0,0019)	*3,0000* *(0,0010)*	3,0000 (0,0027)	*3,0000* *(0,0005)*
Credit_coop	114,0000 (6,0000)	42,0000 (2,0000)	6,2966 (0,2990)	*5,9874* *(0,2429)*	5,6899 (0,2801)	*5,6920* *(0,2245)*	5,4772 (0,1837)	*5,4958* *(0,2128)*	5,3535 (0,2650)	*5,2990* *(0,1907)*
Credit_g	98,0009 (0,9868)	78,0015 (0,3214)	7,7909 (0,0407)	*7,7612* *(0,0540)*	8,2722 (0,0999)	*8,3848* *(0,0141)*	8,6670 (0,0622)	*8,3595* *(0,0608)*	8,8927 (0,0903)	*8,2120* *(0,0564)*
Credit_bank1	154,0000 (5,0000)	81,0000 (2,0000)	7,4939 (0,2928)	*6,9930* *(0,2121)*	7,0936 (0,2642)	*6,3888* *(0,2828)*	7,6993 (0,2307)	*6,7588* *(0,2642)*	7,5894 (0,1880)	*6,3972* *(0,1915)*
Credit_bank2	126,0000 (4,0000)	59,0000 (3,0001)	8,4994 (0,1907)	*6,4415* *(0,1814)*	8,2951 (0,2121)	*6,3990* *(0,2099)*	6,4926 (0,3145)	*6,6904* *(0,3021)*	6,2801 (0,3536)	*5,9901* *(0,3112)*
Diabetes	26,0017 (0,5129)	9,0163 (0,3223)	4,3517 (0,4218)	*4,2658* *(0,2003)*	4,4488 (0,0327)	*4,2154* *(0,2098)*	4,2687 (0,3327)	*4,2513* *(0,2914)*	4,3419 (0,2759)	*4,1371* *(0,1647)*
Drugte	17,1001 (0,0425)	14,0204 (0,0171)	7,3498 (0,0707)	*8,1167* *(0,2440)*	7,2481 (0,2121)	*7,9860* *(0,2644)*	8,6081 (0,0221)	*7,7944* *(0,3127)*	7,7459 (0,3536)	*7,2002* *(0,3466)*
DrugY	12,0016 (0,0039)	10,0162 (0,0401)	8,5498 (0,0707)	*8,4304* *(0,0319)*	7,8771 (0,0348)	*8,3984* *(0,0487)*	8,4994 (0,0914)	*8,2753* *(0,0497)*	8,2377 (0,0636)	*7,7301* *(0,0168)*
Heart_c	19,0032 (0,2711)	22,0191 (0,2251)	4,7833 (0,0565)	*3,7380* *(0,0512)*	5,0079 (0,0119)	*3,8452* *(0,0201)*	5,2962 (0,0282)	*3,6829* *(0,0371)*	4,7958 (0,0182)	*3,5908* *(0,0129)*
Heart_Statlog	18,0093 (0,2720)	24,0152 (0, 2670)	4,8836 (0,0445)	*3,8806* *(0,0400)*	4,6957 (0,0092)	*3,5804* *(0,0039)*	5,0906 (0,0078)	*3,7929* *(0,0280)*	4,0496 (0,0407)	*3,3468* *(0,0316)*
Wines	7,1002 (0,0105)	5,0019 (0,0726)	4,3358 (0,0750)	*5,3102* *(0,0810)*	4,3954 (0,0282)	*5,1066* *(0,0221)*	4,6989 (0,0141)	*4,4981* *(0,0106)*	4,1496 (0,0707)	*4,1144* *(0,0481)*
Zoo	9,0101 (0,0190)	8,0018 (0,0094)	6,9992 (0,0141)	*6,9970* *(0,0197)*	7,1000 (0,0011)	*6,9951* *(0,0015)*	7,0281 (0,0145)	*6,8745* *(0,0106)*	6,9832 (0,0219)	*6,5422* *(0,0156)*

To verify the performance of this method, we used twelve databases from the UCI Repository, one database of a savings and credit cooperative of the Financial System of Ecuador and two databases from two Banks of Ecuador. We run 20 independent executions of each method. The LVQ network is made up of 30 neurons.

Table 1 shows the classification accuracy of the different methods applied on the mentioned databases. We considered the precision of the set of rules that use fuzzy and nominal variables in the antecedent, indicating also their respective standard deviation. Table 2 shows the results of the average number of rules obtained in each method. The antecedent of the rules is formed by a nominal and/or fuzzy variables. The table also includes the information regarding the respective standard deviation.

To test the hypothesis that fuzzy models are better than those that do not use fuzzy variables, difference-in-mean tests were carried out. The tests verified that fuzzy models present in most cases better precision than PART method, and slightly higher than C4.5 method. However, if we focus on the number of rules generated to achieve such precision, we detect that fuzzy methods use substantially fewer number of rules (on average), than C4.5 or PART. This results reinforces our claim that our method simplifies the interpretation of results.

5 Conclusions

A new method of fuzzy classification rules has been presented, using fuzzy and/or nominal variables. This method uses a neural network with supervised learning (LVQ), and a variable population metaheuristic technique (varPSO), using a voting criterion for the movement of particles, obtaining fuzzy and/or nominal attributes in the background of the rule.

From the numerical experiments, it can be seen that the fuzzy models in all cases obtain a simpler model, achieving significant lower number of rules. The reduction in rules cardinality results in better interpretability. In addition, our model improves accuracy.

In future work, we will model the consequent of the rule as a fuzzy variable. In this way, the result will be expressed as membership function, which could provide additional information related to the credit risk involved in the acceptance of a given credit application.

References

Aggarwal, C.: Data Mining: The Textbook. Springer, New Delhi (2015)

Chen, C.-C.: A PSO-based method for extracting fuzzy rules directly from numerical data. Cybern. Syst. Int. J. **37**, 707–723 (2006)

Eibe, F., Witten, I.H.: Generating accurate rule sets without global optimization. In: Fifteenth International Conference on Machine Learning, pp. 144–151 (1998)

Ganeshkumar, P., Rani, C., Deepa, N.: Formation of fuzzy IF-THEN rules and membership function using enhanced Particle Swarm Optimization. Int. J. Uncertainty Fuzziness Knowl. Based Syst. **21**(1), 103–126 (2013)

Hasperué, W., Lanzarini, L.: Dynamic self-organizing maps. A new strategy to upgrade topology preservation. In: XXXI Conf. Latinoamericana de Informática, CLEI 2005 (2005)

Kennedy, J., Eberhart, R.: Particle swarm optimization. In: Proceedings of IEEE International Conference on Neural Networks, vol. 4, pp. 1942–1948 (1995)

Kohonen, T.: Self-organizing Maps. Springer Series in Information Sciences, vol. 30. Springer, Heidelberg (2012)

Lanzarini, L., Villa Monte, A., Aquino, G., De Giusti, A.: Obtaining classification rules using lvqPSO. In: Advances in Swarm and Computational Intelligence. LNCS, vol. 6433, pp. 183–193. Springer, Heidelberg (2015a)

Lanzarini, L., Villa-Monte, A., Bariviera, A., Jimbo, P.: Obtaining classification rules using LVQ +PSO: an application to credit Risk, pp. 383–391. Springer (2015b)

Lanzarini, L., Villa-Monte, A., Bariviera, A., Jimbo, P.: Simplifying credit scoring rules using LVQ+PSO. Kybernetes **46**, 7–16 (2017)

Quinlan, J.R.: C4.5: Programs for Machine Learning. Morgan Kaufmann Publishers, San Mateo (1993)

UC Irvine Machine Learning Repository. http://archive.ics.uci.edu/ml/

Venturini, G.: SIA: a supervised inductive algorithm with genetic search for learning attributes based concepts. In: Brazdil, P. (ed.) Machine Learning: ECML-93. LNCS, vol. 667, pp. 280–296. Springer, Heidelberg (1993)

Ziqiang, W., Xia, S., Dexian, Z.: A PSO-based classification rule mining algorithm. In: Advanced Intelligent Computing Theories and Applications. LNCS, vol. 4682, pp. 377–384. Springer (2007)

Witten, I., Eibe, F., Hall, M., Pal, C.: Data Mining - Practical Machine Learning Tools and Techniques, 4th edn. Morgan Kaufmann (2016). ISBN 9780128042915

Yager, R.R., Zadeh, L.A. (eds.): An introduction to fuzzy logic applications in intelligent systems. In: The Springer International Series in Engineering and Computer Science, vol. 165. Springer (2012)

Zadeh, L.A.: Fuzzy sets. Inf. Control **8**(3), 338–353 (1965)

Revisiting Data Augmentation for Rotational Invariance in Convolutional Neural Networks

Facundo Quiroga[1,2], Franco Ronchetti[1,3], Laura Lanzarini[1], and Aurelio F. Bariviera[4(✉)]

[1] Facultad de Informática, Instituto de Investigación en Informática LIDI (Centro CICPBA), Universidad Nacional de La Plata, La Plata, Argentina
{fquiroga, fronchetti, laural}@lidi.info.unlp.edu.ar
[2] Becario de postgrado de la Universidad Nacional de La Plata, La Plata, Argentina
[3] Becario postdoctoral de la Universidad Nacional de La Plata, La Plata, Argentina
[4] Business Management Department, Universitat Rovira i Virgili, Tarragona, Spain
aurelio.fernandez@urv.cat

Abstract. Convolutional Neural Networks (CNN) offer state of the art performance in various computer vision tasks. Many of those tasks require different subtypes of affine invariances (scale, rotational, translational) to image transformations. Convolutional layers are translation equivariant by design, but in their basic form lack invariances. In this work we investigate how best to include rotational invariance in a CNN for image classification. Our experiments show that networks trained with data augmentation alone can classify rotated images nearly as well as in the normal unrotated case; this increase in representational power comes only at the cost of training time. We also compare data augmentation versus two modified CNN models for achieving rotational invariance or equivariance, Spatial Transformer Networks and Group Equivariant CNNs, finding no significant accuracy increase with these specialized methods. In the case of data augmented networks, we also analyze which layers help the network to encode the rotational invariance, which is important for understanding its limitations and how to best retrain a network with data augmentation to achieve invariance to rotation.

Keywords: Neural Networks · Convolutional networks · Rotational invariance · Data augmentation · Spatial Transformer Networks · Group Equivariant Convolutional Networks · MNIST · CIFAR10

1 Introduction

Convolutional Neural Networks (CNNs) currently provide state of the art results for most computer vision applications (Dieleman et al. 2016). Convolutional layers learn the parameters of a set of FIR filters. Each of these filters can be seen as a weight-tied version of a traditional feedforward layer. The weight-tying is performed in such a way that the resulting operation exactly matches the convolution operation.

J. C. Ferrer-Comalat et al. (Eds.): MS-18 2018, AISC 894, pp. 127–141, 2020.
https://doi.org/10.1007/978-3-030-15413-4_10

While feedforward networks are very expressive and can approximate any smooth function given enough parameters, a consequence of the weight tying scheme is that convolutional layers do not have this property. In particular, traditional CNNs cannot deal with objects in domains where they appear naturally rotated in arbitrary orientations, such as texture recognition (Marcos et al. 2016), handshape recognition (Quiroga et al. 2017), or galaxy classification (Dieleman et al. 2015).

Dealing with rotations (or, in general, with geometric transformations) requires the network to be invariant or equivariant to those transformations. A network $\mathbf{f}$ is invariant to a transformation φ if transforming the input to the network $\mathbf{x}$ with φ does not change the output of the network, that is, we have $\mathbf{f}(\varphi(\mathbf{x})) = \mathbf{f}(\mathbf{x})$ for all $\mathbf{x}$. A network is equivariant to a transformation φ if its output changes *predictably* when the input is transformed by φ. Formally, it is equivariant if there exists a smooth function φ' such that for all $\mathbf{x}$, we have $\mathbf{f}(\varphi(\mathbf{x})) = \varphi'(\mathbf{f}(\mathbf{x}))$ (Dieleman et al. 2016).

Depending on the application, invariance and/or equivariance may be required in different layers of the network.

While traditional CNNs are translation equivariant by design (Dieleman et al. 2016), they are neither invariant nor equivariant to other types of transformations in usual training/usage scenarios. There are two basic schemes for providing rotation invariance to a network: augmenting the data or the model.

Data-augmentation is a very common method for achieving invariance to geometric transformations of the input and improving generalization accuracy. Invariance and equivariance to rotations via data augmentation has been studied for Deep Restricted Boltzmann Machines (Larochelle et al. 2007) as well as HOGs and CNNs (Lenc and Vedaldi 2015). These results show evidence in favour of the hypothesis that traditional CNNs can learn automatically equivariant and invariant representations by applying transformations to their input. However, these networks require bigger computational budget in the training period, since the transformation space of the inputs must be explored by transforming them.

Other approaches modify the model or architecture instead of the input to achieve invariance. For example, some modified models employ rotation invariant filters, or pool multiple predictions made with rotated versions of the input images. However, it is unclear how and whether these modifications improve traditional CNNs trained with data augmentation, both in terms of efficiency and representational power. Furthermore, the mechanisms by which traditional CNNs achieve invariances to rotation are still poorly understood. How best to augment data to achieve rotational invariance is also unclear.

This paper compares modified CNNs models with data augmentation techniques for achieving rotational invariance for image classification. We perform experiments with various well understood datasets (MNIST, CIFAR10), and provide evidence for the fact that despite clever CNNs modifications, data augmentation is still necessary with the new models and paired with traditional CNNs can provide similar performance while remaining simpler to train and understand.

2 Review of Convolutional Neural Networks Models with Rotational Invariance

In this subsection we review the literature on modified CNN models for rotation invariance.

Many modifications for CNNs have been proposed to provide rotational invariance (or equivariance) (Jaderberg et al. 2015, Laptev et al. 2016, Gens and Domingos 2014, Wu et al. 2015, Marcos et al. 2016, Dieleman et al. 2016, Cohen and Welling 2016, Cohen and Welling 2017, Zhou et al. 2017).

Some researchers claim that for classification it is usual to prefer that the lower layers of the network encode equivariances, so that multiple representations of the input can coexist, and the upper layers of the network encode invariances, so that they can collapse those multiple representations in a useful fashion (Lenc and Vedaldi 2015). In this way, we can make the network learn all the different orientations of the object as separate entities, and then map all those representations to a single class label.

Alternatively, we can add an explicit image reorientation scheme that is applied to the image before passing it as an input to the network. In this way, the network can ignore the rotation of the object and learn a representation in a unique, canonical orientation, which simplifies the network design. However, this requires an additional model that can predict the orientation of the object.

The first approach puts the invariance near the output; the second puts it near the input by making the input layer rotation invariant. Moreover, for some objects we desire not just whole-image rotation invariance, but also invariance for some of the object's parts. For example, the arms of a person may rotate around the shoulders. It is clear that for these types of problems making the input invariant to global rotations is insufficient.

The following subsection reviews modified CNN models that deal with rotation (in, equi)-variance. We divide them into two groups: those that attempt to deal with the rotation problem globally by **transforming the input image or feature map**, and those that propose to **modify the convolution layer** in some sense to produce equivariant features and, optionally, a way to turn those features into invariant ones.

2.1 Transformation of the Input Image or Feature Map

Spatial Transformer Networks (STN) (Jaderberg et al. 2015) defines a new Spatial Transformer Layer (STL) (Figure) that can learn to rotate input feature maps to a canonical orientation so that subsequent layers can focus on the canonical representation. Actually, STLs can also learn to correct arbitrary affine transformations by employing a sub-network that takes the feature maps as a parameter and outputs a 6-dimensional vector that encodes the affine transformation parameters. The transformation is applied via a differentiable bilinear interpolation operation. While typically the STLs are added as the first layer of the network, the layers are modular and can be added at any point in the network's convolutional pipeline (Fig. 1).

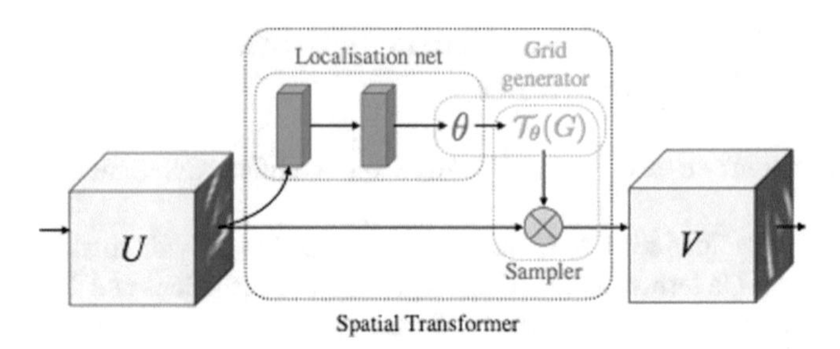

Fig. 1. Architecture of a Spatial Transformer Layer from (Jaderberg et al. 2015). The layer transform feature maps **U** to feature maps **V** by applying an affine transform **T**. The parameters of the affine transform $\boldsymbol{\theta}$ are predicted by a localization network.

Deep Symmetry Networks (DSN) (Gens and Domingos 2014) also transforms the image *prior* to convolution and max-pools the results, but adds an iterative optimization procedure over the 6-dimensional space of affine transformation to find a *transformation* that maximally activates the filter, mixing of ideas from TIP and STN. In spirit, their approach is similar to the STN approach but the optimization procedure is less elegant than the jointly trained localization network and could be seen as well as a form of data augmentation. They compare against pure data augmentation in the MNIST and NORB datasets and find that while DSNs has better performance when training with less than 10000 samples, data augmentation achieves the same performance at that point.

Transformation-Invariant Pooling (TIP) (Laptev et al. 2016) proposes to define a set of transformations $\boldsymbol{\Phi} = \{\varphi[\mathbf{1}], \ldots, \varphi[\mathbf{n}]\}$ to which the network must have invariance, and then train a siamese network with a subnetwork $\mathbf{N[i]}$ for each transformation $\varphi[\mathbf{i}]$. The input to the i-th subnetwork is $\varphi[\mathbf{i}](\mathbf{x})$, that is, subnetworks share parameters but each is fed with an input transformed in a different way. A max-pooling operation is performed on the vector of outputs of the siamese network for each class. In this way the output of the network is still a vector of probabilities, one for each class. This pooling operation is crucial since it provides the invariance needed; before that operation, the representation would be (at most) equivariant.

The set of transformations $\boldsymbol{\Phi}$ is user-defined and can include a set of fixed rotations; the authors show that whenever $\boldsymbol{\Phi}$ forms a group then the siamese network is guaranteed to be invariant to $\boldsymbol{\Phi}$ (assuming the input images are only affected by that set of transformations, else the network will be *approximately* invariant). TIP can be viewed as a form of test-time data augmentation, that prepares the network by data-augmentating the training of the feature extraction part of the network.

2.2 Modifications of the Convolution Layer

Flip-Rotate-Pooling Convolutions (FRPC) (Wu et al. 2015) extends the convolution layer by rotating the filter instead of the image. In this way, a oriented convolution generates additional feature maps by rotating traditional convolutional filters in **n*r** fixed orientations (**n*r** is a parameter). Then, max-pooling along the channel dimension is applied to the responses of the **n*r** orientations so that a single feature map results.

No comparison to data augmentation approaches was performed. While the number of parameters of the layer is not increased, the run-time memory and computational requirements of the layer are multiplied by **n*r**, although the number of parameters for the full network can actually be reduced since the filters are more expressive.

The same approach is used in (Marcos et al. 2016) for texture classification; they do perform comparisons with data augmentation but only for 20 samples. Given that the rotation prior is very strong in texture datasets, this is an unfair comparison.

Exploiting Cyclic Symmetry in CNNs (Dieleman et al. 2016) presents a method similar to FPRC alongside variants that provide equivariance as well as invariance. *Oriented response networks (ORN)* (Zhou et al. 2017) are related to FRPC, but also introduce an ORAlign layer that instead of pooling the set of features maps generated by a rotating filter reorders them in a SIFT-inspired fashion which also provides invariance.

Dynamic Routing Between Capsules (Sabour et al. 2017) presents a model are units of neurons designed to mimic the action of cortical columns. Capsules are designed to be invariant to complicated transformations of the input. Their outputs are merged at the deepest layer, and so are only invariant to global transformation.

Group Equivariant Convolutional Networks (GCNN) and *Steerable CNNs* (Cohen and Welling 2016, Cohen and Welling 2017) also use the same basic methods but provide a more formal theory for guaranteeing the equivariance of the intermediate representations by viewing the set of transformations of the filters as a group. *Learning Steerable Filters for Rotation Equivariant CNNs* (Weiler et al. 2018) also employs the same approach. *Spherical CNNs* (Cohen et al. 2018) extend this approach to the 3D case.

In particular, Group CNNs (Cohen and Welling 2016) adds an additional *rotation* dimension to the convolutional layer. This dimension allows to compute rotated versions of the feature maps in 0°, 90°, 180° and 270° orientations, as well as their corresponding horizontally flipped versions. The first convolution *lifts* the image channels by adding this dimension; further convolutions compute the convolution across all channels and rotations, so that the filter parameters are of size **Channels*Rotations*H*W**, where H and W are the height and width of the filter. The bias term is the same for all rotation dimensions. To *return* to the normal representation of feature maps, the rotation dimension is reduced via a max operation, obtaining invariance to rotation (Fig. 2).

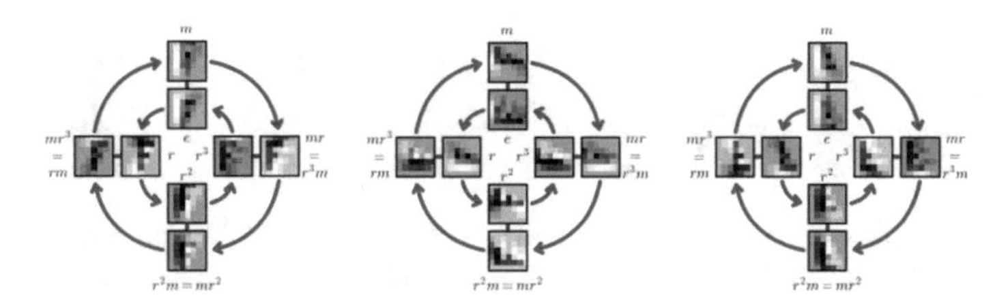

Fig. 2. Transformations of a filter made by a Group CNN from (Cohen and Welling 2016). Filters are rotated and flipped and then applied to the feature maps. The rotation and flip operations form an algebraic group.

We note that all previous models are variations of the same strategy: augment convolutions with equivariance to a finite set of transformations by rotating or adapting the filters, then provide a method to collapse the equivariance into an invariance before the final classification.

Deformable Convolutional Networks (Dai et al. 2017) learns filters of arbitrary shape. Each position of the filter can be arbitrarily spatially translated, via a mapping that is learned. Similarly to STN, the employ a differentiable bilinear filter strategy to sample from the input feature map. While not restricted to rotation, this approach is more general than even STN since the transformation to be learned is not limited by any affine or other priors; hence we do not consider it for the experiments.

3 Experiments

We performed two types of experiments understand data augmentation for rotational invariance and compare it with other methods. We used the MNIST and CIFAR10 datasets (Fig. 3) in our experiments (Le Cunn et al. 1998, Krizhevsky et al. 2009) because these are well known and the behavior of common networks such as those tested here is better understood than for other datasets, and cover both synthetic grayscale images and RGB natural images.

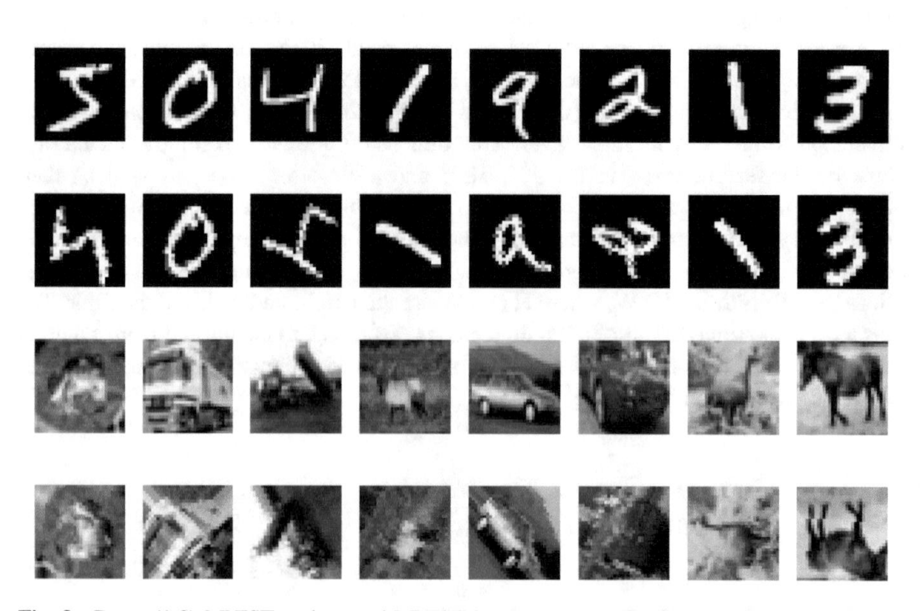

Fig. 3. Rows (1,2) MNIST and rotated MNIST images. (3,4) CIFAR10 and rotated CIFAR10 images. Note that for CIFAR10 some images are slightly cropped by the rotation procedure, while for MNIST the cropping is negligible due to the black border.

The data augmentation we employed consists of rotating the input image by a random angle in the range [0°, 360°]. For the MNIST dataset, previous works employ the MNISTrot version in which images are rotated in 8 fixed angles. While using a discrete set of rotations allows some models to learn a fixed set of filters for guaranteed equivariance, we chose to use a continuous set of rotations because it better reflects real-world usage of the methods. We tested only global rotations; i.e., rotations of the whole image around the center.

In all experiments networks were trained until convergence by monitoring the test accuracy, using the ADAM optimization algorithm with a learning rate of 0.0001 and 1E-9 weight decay for all non-bias terms.

3.1 Network Models

We employed a simple convolutional network we will call **SimpleConv**, that while it clearly does not provide state of the art results, is easy to understand and well. The simple convolutional network for is defined as: Conv1(F) - Conv2(F) – MaxPool1 (2 × 2) – Conv3(F*2) – Conv4(F*2) – MaxPool2(2 × 2) – Conv5(F*4) – FC1(D) – Relu – BatchNorm – FC(10), where all convolutional filters are 3 × 3, and there is a ReLU activation function after each convolutional and fully connected . F is the number of feature maps of the convolutional layers and D the hidden neurons of the fully connected layer. For MNIST, we set F = 32 and D = 64, while for CIFAR10 F = 64 and D = 128, matching the original Group CNN implementation (see below).

To test the importance of the Dense layers in providing invariance to rotated samples, we also experimented with an **AllConvolutional** network. This model uses just convolutions and pooling as building blocks, which is of interest to our. The AllConv network is defined as: Conv1(F) – Conv2(F) – Conv3(F, stride = 2 × 2) - Conv4(F*2) – Conv5(F*2) – Conv6(F*2, stride = 2 × 2) – Conv7(F*2) – Conv8(F*2, 1 × 1) - ConvClass(10, 1 × 1) – GlobalAveragePooling(10). Again, all convolutional filters are 3 × 3, and we place ReLUs after convolutions. For MNIST, F = 16 while for CIFAR10, F = 96, again matching the original Group CNN implementation.

Then, we chose a model from the two groups described in Sect. 2 - transformation of the input image and transformation of the filters – that also correspond to the alternative strategies of putting the invariance near the input or near the output.

As a representative of the first group we added a **Spatial Transformer Layer** (Jaderberg et al. 2015) to the convolutional network to reorient the image before the network classifies it; the resulting networks are named **SimpleConvSTN** and **AllConvolutional STN**. To keep comparisons fair, we modified the localization and affine matrix to restrict transformations to rotations. The localization network consists of a simple CNN with layers: Conv1(16, 7 × 7) – MaxPool1(2 × 2) – ReLU() – Conv2(16, 5 × 5) – MaxPool2(2 × 2) – ReLU() – FC(32).

From the modified convolutional methods, we chose **Group CNNs** (Cohen and Welling 2016). The resulting networks are named **SimpleGConv** and **AllGConvolution**, where we simply replaced normal convolutions for group convolutions, and we added a pooling operation before the classification layers to provide the required invariance to rotation. Group CNNs has 4 times the feature maps as a regular CNN network; to compensate, and as a compromise, we reduced the parameters by half.

3.2 Data Augmentation with Traditional Networks

First we measured the performance of a SimpleConv and AllConvolutional with and without data augmentation to obtain a baseline for both base methods.

We trained two instances of each model; one with the normal dataset, and the other with a data-augmented version. We then tested each instance of each model with the test set of the normal and data-augmented variant. Figure 4 shows the results of the experiments for each model/dataset combination.

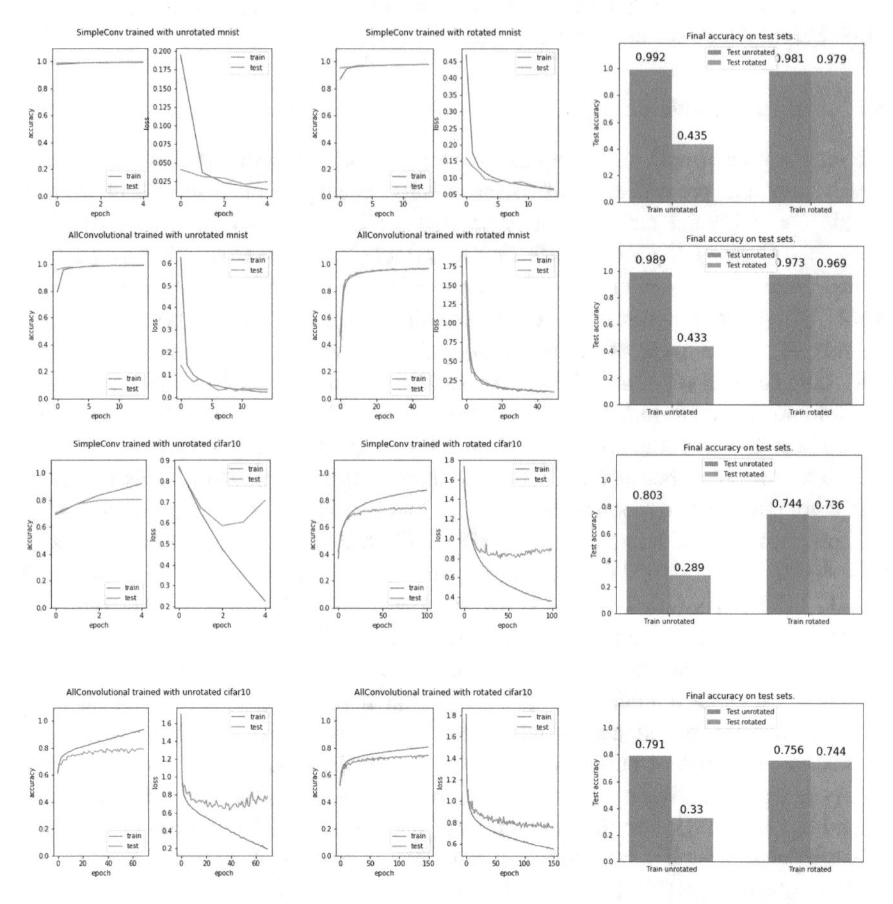

Fig. 4. Rows (1,2) SimpleConv and AllConvolutional with MNIST. (3,4) SimpleConv and AllConvolutional with CIFAR10. Left Accuracy and loss for each training epoch, on training and test set. Middle Same as left, but with a rotated training set. Right Final test set accuracies for the two models trained and the two variations of the dataset (unrotated, rotated).

On MNIST, we can observe that while networks trained with a rotated dataset see a drop in their accuracies of 1%–2% on the unrotated test set, the performance drop for networks trained with the unrotated dataset and tested with the rotated dataset fare much worse ($\sim$55% drop in accuracy). It is surprising that the drop in the first case is so low, specially given that the number of parameters is the same for both networks. This may indicate a redundancy in the filters of the unrotated model. Also, it would seem that the network trained on the unrotated dataset still performs at a $\sim$40% accuracy level, four times more than expected by chance (10%). This is partly because some of the samples are naturally invariant to rotations such as the images for the number 0 or invariant to some rotations such as the numbers 1 or 8, and because some of the learned features must be naturally invariant to rotation as well.

In the case of CIFAR10 the results show a similar situation, although the drop in accuracy from unrotated to rotated is larger, possibly because the dataset possesses less natural invariances than MNIST. Note that to reduce the burden of computations the number of training epochs on CIFAR10 with AllConvolutional was reduced, achieving $\sim$80% accuracy instead of 91% as in the authors original experiments (Jaderberg et al. 2015).

Still, it is surprising that the AllConvolutional networks can learn the rotated MNIST dataset so well, since convolutions are neither invariant nor equivariant to rotation. This points to the fact that the set of filters learned by the network possibly self-organize during learning to obtain a set of filters that can represent all the rotated variations of the object.

3.3 Comparison with STN and Group CNN

We ran the same experiment as before but using the modified versions of the network. Figure 5 shows the results of the STLs versions of the networks.

We can see that in all cases the performance of the unrotated model on the rotated dataset is much lower than the original; this is expected since the STL needs data augmentation during training. However, the STL model did not perform noticeable better than the normal data-augmentated models (Fig. 4).

Figure 6 shows as well the results on MNIST and CIFAR10 of the Group CNN models. Similarly to the STN case, the performance is not increased with Group CNN: however, in the case of the AllGConvolutional network we see that the performance of the model trained with the unrotated dataset on the rotated dataset is much greater (+0.1) than for other models, while the same does not happen with the SimpleGConv network. This is possibly due to the fact that the superior representation capacity of the fully convolutional layers can compensate for absence of good filters, while in the AllGConvolutional case there is more pressure on the convolutional layers to learn good representations.

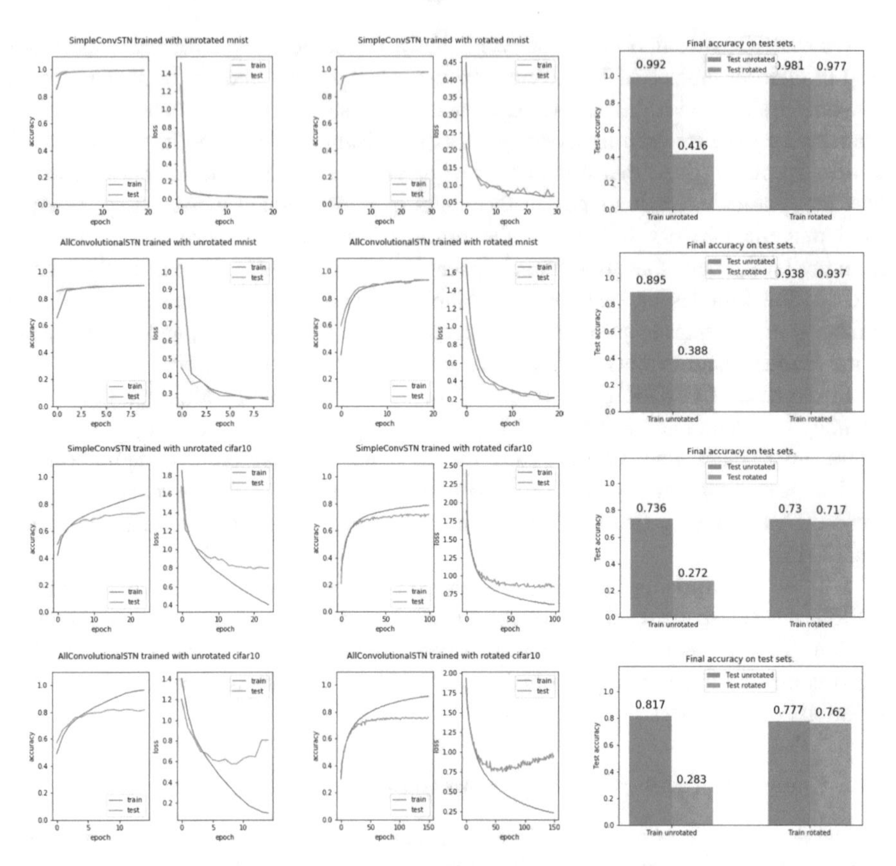

Fig. 5. Rows (1,2) SimpleConvSTN and AllConvolutionalSTN with MNIST. (3,4) SimpleConvSTN and AllConvolutionalSTN with CIFAR10. Left Accuracy and loss for each training epoch, on training and test set. Middle Same as left, but with a rotated training set. Right Final test set accuracies for the two models trained and the two variations of the dataset (unrotated, rotated)

3.4 Retraining for Rotational Invariance

While Sect. 3.2 seems to point to the fact that there is small difference in accuracy when using specialized models versus data augmentation, it could be argued that specialized models can be more efficient when training. Alternatively, if training time is a limiting factor, we could use a pretrained network and retrain some of its layers to achieve invariance. However, a priori it is not clear whether the whole network can/needs to be retrained, or if only some parts of it need to adapt to the rotated examples.

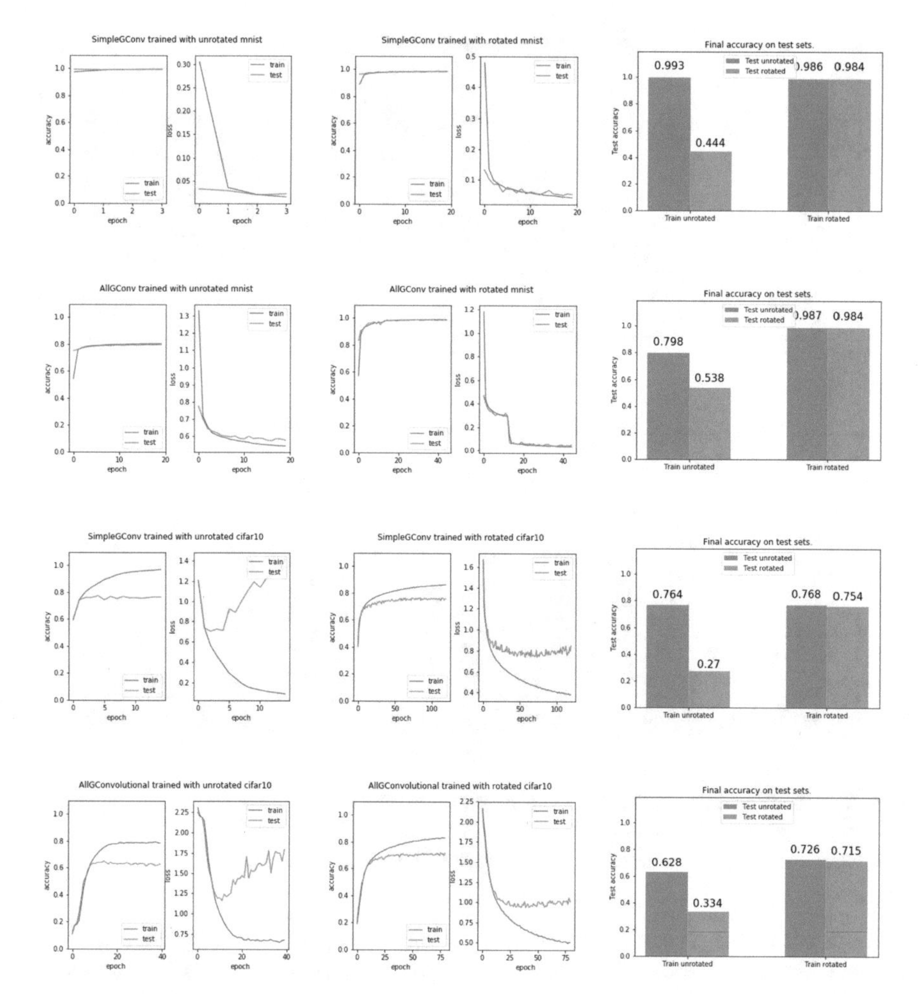

Fig. 6. Rows (1,2) SimpleGConv and AllGConvolutional with MNIST. (3,4) SimpleGConv and AllGConvolutional with CIFAR10. Left Accuracy and loss for each training epoch, on training and test set. Middle Same as left, but with a rotated training set. Right Final test set accuracies for the two models trained and the two variations of the dataset (unrotated, rotated)

To analyze which layers are amenable to retraining for rotation invariance purposes, we train a base model with an unrotated dataset, then make a copy of the network and retrain parts of it (or all of it) to assess which layers can be retrained.

As Fig. 7 shows, retraining for rotation invariance shows a similar trend to retraining for transfer learning; higher layers with more high-level features are a better target for retraining individually since they are closer to the output layers and can affect them more. The fact that retraining the penultimate layers can bring performance back means that there's a redundancy of information in previous layers since rotated

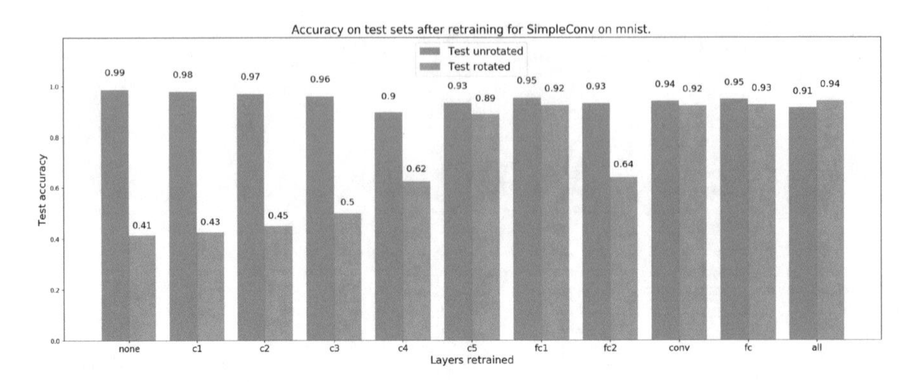

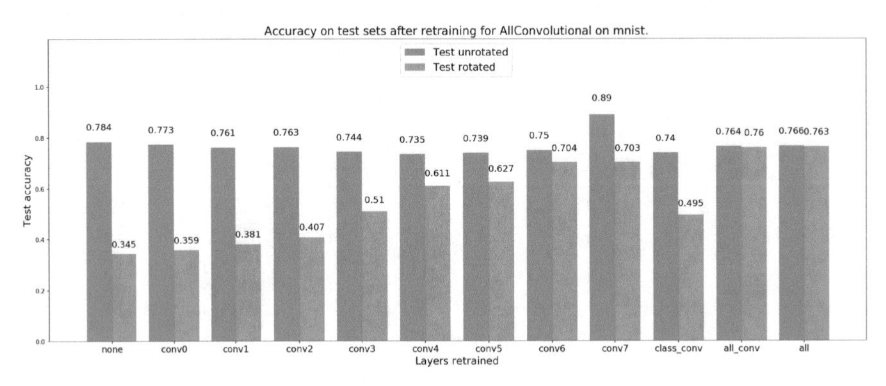

Fig. 7. Retraining experiments for SimpleConv and AllConvolutional on MNIST. Accuracy on the test sets after retraining subsets of layers of the network. The labels "conf"/"all_conv" and "fc" mean that all convolutional or fully connected layers were retrained.

versions of the images can be reconstructed from the output of the first layers. Since the original networks were trained with unrotated examples, this means that either that network naturally learns equivariant filters, or that equivariance in filters is not so important for classifying rotated objects.

However, it is surprising that retraining the final layer in both cases leads to a reversal of this situations: the performance obtained is less than when retraining other layers. In the case of the SimpleConv network, this is possibly due to the action of previous fc layer collapsing the equivariances before the final layer can translate them to a decision; that is, the fc1 layers must be loosing *some* information. In the case of the AllConvolutional network, the class_conv layer performs a simple 1×1 convolution to collapse all feature maps to 10, and so probably cannot recapture the invariances, while the retraining the previous layer with many more 3×3 convolutions can.

Figure 8 shows the results of the same experiment on CIFAR10. The results are similar to those of MNIST, except for the lower general accuracy given the difficulty of CIFAR10.

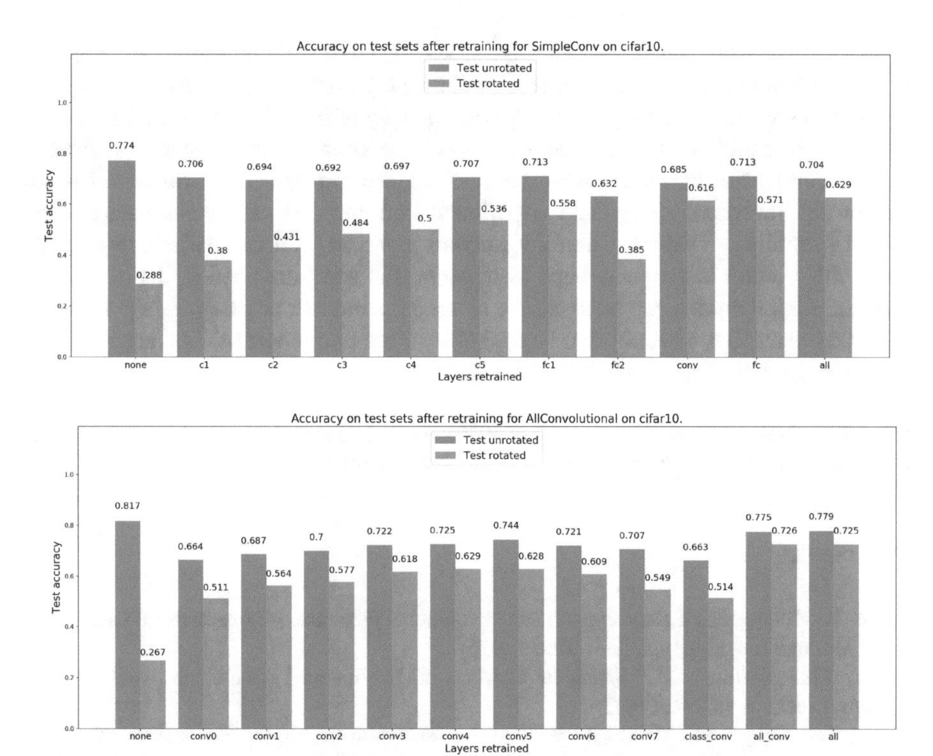

Fig. 8. Retraining experiments for SimpleConv and AllConvolutional on CIFAR10. Accuracy on the test sets after retraining subsets of layers of the network. The labels “conf”/”all_conv” and “fc” mean that all convolutional or fully connected layers were retrained.

4 Conclusions

Rotational invariance is a desired property for many applications in image classification. Data augmentation is a simple way of training CNNs models, which currently hold the state of the art, to achieve invariance. There are many modified CNNs models that attempt to make this task easier.

We compared data augmentation with modified models, maintaining the same number of parameters in each case. While data augmentation requires more training, it can reach similar accuracies as other methods. Furthermore, the test time is not affected by additional localization networks or convolutions.

We also performed retraining experiments with data augmentation to shed some light on how and where can network learn rotational invariance. By retraining layers separately, we found that some invariance can be achieved added in every layer, although the ones nearest the end of the network, whether fully connected or convolutional, are much better at gaining invariance. This finding reinforces the notion that lower layers of networks learn redundant filters and so can be capitalized for other tasks (rotation invariant tasks in this case) and also the notion that invariance should be added at the end of the network if possible.

We believe more can be learned about CNNs by studying their learned invariances. Possible extensions of this work include deliberately introducing invariances early or late in the network to see how they affect the capacity of the network to learn rotations. It would be useful as well to compare the convolutional filters learned to identify equivariances between their outputs, as well as see via their activations which are rotationally invariant. Systematic experimentation on datasets where samples are rotated naturally, as well as naturally rotation invariant, and comparison to datasets where the rotation is synthetic such as those in this work are needed. To reduce the experimentation burden, we have chosen to keep the number of parameters constant in the experiments, but it would also be desirable to see the impact of data augmentation when the number of parameters is constrained.

Acknowledgements. The authors thank NVIDIA for their donation of a state of the art GPU which facilitated the execution of the experiments in this article.

References

Cohen, T., Welling, M.: Group equivariant convolutional networks. In: International Conference on Machine Learning, pp. 2990–2999 (2016)

Cohen, T.S., Welling, M.: Steerable CNNs. In: International Conference on Learning Representations (ICLR) (2017)

Cohen, T.S., Geiger, M., Köhler, J., Welling, M.: Spherical CNNs. In: Proceedings of the 6th International Conference on Learning Representations (ICLR) (2018)

Dai, J., Qi, H., Xiong, Y., Li, Y., Zhang, G., Hu, H., Wei, Y.: Deformable convolutional networks. In: Proceedings of the 2018 International Conference on Computer Vision (2017)

Dieleman, S., De Fauw, J., Kavukcuoglu, K.: Exploiting cyclic symmetry in convolutional neural networks. In: ICML 2016 Proceedings of the 33rd International Conference on International Conference on Machine Learning, vol. 48, pp. 1889–1898 (2016)

Dieleman, S., Willett, K.W., Dambre, J.: Rotation invariant convolutional neural networks for galaxy morphology prediction. Mon. Not. R. Astron. Soc. **450**(2), 1441–1459. arXiv:1503.07077 (2015)

Gens, R., Domingos, P.M.: Deep symmetry networks. Adv. Neural. Inf. Process. Syst. **27**, 2537–2545 (2014)

Jaderberg, M., Simonyan, K., Zisserman, A.: Spatial transformer networks. In: Advances in Neural Information Processing Systems (NIPS 2015), pp. 2017–2025 (2015)

Krizhevsky, A., Hinton, G.: Learning multiple layers of features from tiny images. Technical report, University of Toronto, vol. 1, no. 4, p. 7 (2009)

Laptev, D., Savinov, N., Buhmann, J. M., Pollefeys, M.: TI-POOLING: transformation-invariant pooling for feature learning in convolutional neural networks. In: International Conference on Computer Vision and Pattern Recognition (CVPR) 2016, pp. 289–297. IEEE (2016)

Larochelle, H., Erhan, D., Courville, A., Bergstra, J., Bengio, Y.: An empirical evaluation of deep architectures on problems with many factors of variation. In: Proceedings of the 24th International Conference on Machine Learning, ICML 2007, pp. 473–480. ACM, New York (2007)

LeCun, Y., Bottou, L., Bengio, Y., Haffner, P.: Gradient-based learning applied to document recognition. Proc. IEEE **86**(11), 2278–2324 (1998)

Lenc, K., Vedaldi, A.: Understanding image representations by measuring their equivariance and equivalence. In: 2015 IEEE Conference on Computer Vision and Pattern Recognition (CVPR), pp. 991–999, June 2015

Marcos, D., Volpi, M., Tuia, D.: Learning rotation invariant convolutional filters for texture classification. In: 2016 23rd International Conference on Pattern Recognition (ICPR), pp. 2012–2017. IEEE, December 2016

Quiroga, F., Antonio, R., Ronchetti, F., Lanzarini, L.C., Rosete, A.: A study of convolutional architectures for handshape recognition applied to sign language. In XXIII Congreso Argentino de Ciencias de la Computación, La Plata (2017)

Sabour, S., Frosst, N., Hinton, G.E.: Dynamic routing between capsules. In: Advances in Neural Information Processing Systems, pp. 3856–3866 (2017)

Weiler, M., Hamprecth, F., Storath, M.: Learning steerable filters for rotation equivariant CNNs. In: The IEEE Conference on Computer Vision and Pattern Recognition (CVPR) (2018)

Wu, F., Hu, P., Kong, D.: Flip-rotate-pooling convolution and split dropout on convolution neural networks for image classification. arXiv preprint arXiv:1507.08754 (2015)

Zhou, Y., Ye, Q., Qiu, Q., Jiao, J.: Oriented response networks. In: 2017 IEEE Conference on Computer Vision and Pattern Recognition (CVPR), pp. 4961–4970. IEEE, July 2017

Leading Universities in Tourism and Hospitality Research: A Bibliometric Overview

Carles Mulet-Forteza[1(✉)], Emilio Mauleón-Méndez[1], José M. Merigó[2,3], and Juanabel Genovart-Balaguer[1]

[1] Department of Business Administration, University of the Balearic Islands, Carretera Valldemossa, km 7.5, 07122 Palma de Mallorca, Spain
{carles.mulet,emilio.mauleon, juanabel.genovart}@uib.es

[2] Department of Management Control and Information Systems, University of Chile, Diagonal Paraguay 257, 8330015 Santiago, Chile
jmerigo@ub.edu

[3] School of Information, Systems and Modelling, Faculty of Engineering and Information Technology, University of Technology Sydney, 81 Broadway, Ultimo, NSW 2007, Australia

Abstract. This study presents an overview of the most influential and productive institutions around the world in the tourism, leisure and hospitality field. The research uses a bibliometric approach to examine all the journals indexed in the Web of Science database in these fields. Results show a strong influence of the USA institutions, although institutions from other countries also achieve an important and significant position in these fields. Results also consider the temporal evolution of the leading universities and the most productive institutions in the tourism, leisure and hospitality field.

Keywords: Bibliometrics · Institutions research · Productivity · Ranking

1 Introduction

Research on tourism, leisure and hospitality fields has increased considerably in recent years, due to the proliferation of journals in these disciplines. While before the year 2000, based on the Web of Science (WoS) database, only 3,672 documents were published on tourism, leisure and hospitality, between 2001 and 2016 this value has increased a 350%. Using several bibliometric indicators, this paper aims to develop an analysis that identifies the most productive and influential institutions in tourism, leisure and hospitality fields. For it, this paper analyze the production of the world's 20 most influential journals in these fields (according to WoS). The paper identifies the 50 institutions with the largest number of documents and citations and performs a dynamic analysis of these results.

This study is of interest to several stakeholders. For example, the editorial boards of journals can identify institutions with growth potential, students can recognize

J. C. Ferrer-Comalat et al. (Eds.): MS-18 2018, AISC 894, pp. 142–152, 2020.
https://doi.org/10.1007/978-3-030-15413-4_11

institutions with international relevance and companies and governments can choose R&D centers that are convenient to finance.

This paper is organized as follows. In Sect. 2 we present a literature review of bibliometric methods in tourism, leisure and hospitality fields. Section presents the methodology used in this paper. Section 4 shows the results obtained and Sect. 5 summarizes the main results of the paper.

2 Literature Review

The term bibliometrics was introduced by Pritchard (1969) as 'the application of mathematical and statistical methods to books and other means of communication'. There are many others definitions of the bibliometrics. For example, Ye et al. (2012) indicate that bibliometrics examines the results of research, including the topics, methods and samples used, while Merigó et al. (2017a, b) indicate that bibliometrics is a research field that quantitatively studies bibliographic material by analyzing a research area and identifying its leading trends.

The bibliometric papers encompass many disciplines, including accounting (Zhong et al. 2016; Merigó and Yang 2017), economics (Danielsen and Delorme 1976; Wagstaff and Culyer 2012), journals (Merigó et al. 2017a, b; Laengle et al. 2017; Valenzuela et al. 2017), management (Podsakoff et al. 2008; Vogel and Güttel 2013), marketing (Kim and McMillan 2008; Samiee and Chabowski 2012; Leung et al. 2017), and tourism and hospitality (McKercher 2005; McKercher 2008; Benckendorff and Zehrer 2013; Figueroa-Domecq et al. 2015; Köseoglu et al. 2015a, b; Ruhanen et al. 2015; Shao-Jie et al. 2015; Cheng 2016; García-Lillo et al. 2016; Omerzel 2016).

We also identify other notable papers in the field of tourism, leisure and hospitality (see, for example, Jogoratnam et al. 2005a, b; Mason and Cameron 2006; McKercher 2008; Goodall 2009; Benckendorff and Zehrer 2013; Shao-Jie et al. 2015; Köseoglu et al. 2015a, b; Ruhanen et al. 2015; Yuan et al. 2015; García-Lillo et al. 2016; Omerzel 2016; and Yuan et al. 2016). Focusing on studies related to institutions, we note that the paper of Jogoratnam et al. (2005a) only analyzed its results for 3 journals; Jogaratnam et al. (2005b) analyzed the contributions made in 11 journals over a period of only 10 years; Yuan et al. (2015) focused only on 21 institutions, and Goodall (2009) analyzed the relationship between the number of publications of an institution and the fact that the chancellor of the institution was an academic of recognized prestige. Mason and Cameron (2006) analyzed the productivity ranges of institutions to evaluate the relationship between productivity and editorial board membership, but only analyzed 23 institutions. So, a research gap remains in this field, because no papers analyze a larger number of journals or a larger number of years like the ones we analyze in the present paper. Therefore, this paper is particularly useful because it identifies the main institutions that have published in the tourism, leisure and hospitality field and enables the evaluation of output performed over the years by these institutions.

3 Methodology

In order to determine the number of institutions to analyze in the present paper, the WoS database has been used. This database is considered the most influential database in the world because it collects high-quality articles recognized by the scientific community (Merigó et al. 2015a), without prejudice to other databases as Scopus or Google Scholar.

We referenced all the journals indexed in the category 'Hospitality, Leisure; Sport and Tourism', removing those related to 'sports'. We observed that there are 20 journals mainly focused on the tourism, leisure and hospitality fields (Table 1). Data collection was performed during the first half of 2015.

Table 1. Leading journals in tourism, leisure and hospitality research in WoS.

Acronym	Journal title
ATR	Annals of Tourism Research
APJTR	Asia Pacific Journal of Tourism Research
CHQ	Cornell Hospitality Quarterly
CIT	Current Issues in Tourism
IJCHM	International Journal of Contemporary Hospitality Management
IJHM	International Journal of Hospitality Management
IJTR	International Journal of Tourism Research
JHLSTE	Journal of Hospitality, Leisure, Sport & Tourism Education
JHTR	Journal of Hospitality & Tourism Research
JLR	Journal of Leisure Research
JST	Journal of Sustainable Tourism
JTCC	Journal of Tourism and Cultural Change
JTR	Journal of Travel Research
JTTM	Journal of Travel & Tourism Marketing
LS	Leisure Sciences
LSt	Leisure Studies
SJHT	Scandinavian Journal of Hospitality and Tourism
TE	Tourism Economics
TG	Tourism Geographies
TM	Tourism Management

This paper considers a wide range of methods to represent the bibliographic data under study. First, the number of publications and citations are considered. According to Ding et al. (2014), this is considered the most popular bibliometric method. Podsakoff et al. (2008) suggested that the number of citations was more significant than the number of documents because citations measured the influence of one researcher. However, other studies chose to use the total number of publications because they measured productivity. In this paper, we combine both indicators, following Merigó et al. (2015a).

We use additional indicators such as the h-index (Hirsch 2005) and the ratio 'citations per publications' (Merigó et al. 2015b).

Refining the information available in the WoS by article, review, note and letter for the journals included in Table 1, we found 12,815 documents. Next, we selected the top 50 institutions that had published the most documents, which reduced the number of documents to 5,612. The h-index of the field of research studied was 87, meaning that 87 papers have received a minimum of 87 citations. We examined each journal, identifying the contributions of authors linked to research centres excluding university systems. From a dynamic perspective, the study is conducted between 1990 and 2014, and includes a five-year analysis. This approach avoids the risk of possible biases caused by particularly productive or/and unproductive years.

Our study also used the VOS viewer software (Van Eck and Waltman 2010) to map the material through bibliographic coupling and co-authorship analysis. Bibliographic coupling occurs when two papers reference a common third paper in their bibliographies. Co-authorship indicates the volume of publications of a set of variables and how they are connected to each other.

4 Results

This section presents the results of the paper.

4.1 The World's 50 Leading Institutions

Table 2 presents a list of the world's 50 leading institutions in tourism, leisure and hospitality fields.

Table 2. The 50 most productive institutions in tourism, leisure and hospitality fields

R	Institution	C	TP	TC	TC/TP	H
1	Hong Kong Pol U	PRC	556	5,993	10.78	36
2	Texas A M U Coll	USA	263	5,831	22.17	39
3	Penn St U	USA	243	3,609	14.85	32
4	Griffith U	AUS	232	3,093	13.33	28
5	U Waterloo	CAN	194	4,016	20.70	36
6	Purdue U	USA	188	2,012	10.70	24
7	U North Carolina	USA	184	2,661	14.46	29
8	U Queensland	AUS	180	1,889	10.49	23
9	U Surrey	UK	154	2,932	19.04	29
10	U Illinois U Ch	USA	147	2,732	18.59	29
11	Virginia Poly Inst	USA	142	2,857	20.12	27
12	Kyung Hee U	RK	140	1,288	9.20	19
13	Temple U	USA	129	1,145	8.88	18
14	Bournemouth U	UK	122	1,188	9.74	17

(continued)

Table 2. (*continued*)

R	Institution	C	TP	TC	TC/TP	H
15	U Central Florida	USA	121	1,483	12.26	21
16	Sejong U	RK	116	1,417	12.22	21
17	Cornell U	USA	111	604	5.44	13
18	Arizona St U	USA	104	2,360	22.69	24
19	U Nevada Las Vegas	USA	103	1,390	13.50	17
20	Clemson U	USA	95	1,654	17.41	23
21	James Cook U	AUS	92	1,107	12.03	17
22	U Waikato	NZ	90	1,160	12.89	17
23	Monash U	AUS	88	1,171	13.31	20
24	Southern Cross U	AUS	85	672	7.91	13
25	U Florida	USA	83	864	10.41	16
26	Victoria U	NZ	78	803	10.29	15
27	U Strathclyde	UK	76	680	8.95	15
28	U Otago	NZ	76	1,010	13.29	16
29	U Illes Balears	SPN	76	972	12.79	19
30	Colorado St U	USA	75	1,121	14.95	20
31	U Georgia	USA	74	1,284	17.35	21
32	Ben Gurion U	ISR	72	892	12.39	15
33	Sun Yat Sen U	PRC	71	528	7.44	7
34	U Nottingham	UK	69	915	13.26	17
35	U Alberta	CAN	69	1,550	22.46	22
36	USA Department Of Agri	USA	67	1,898	28.33	21
37	U South Carolina	USA	66	669	10.14	14
38	USA Forest Service	USA	65	1,883	28.97	21
39	Oxford Brookes U	UK	65	458	7.05	13
40	U New South Wales	AUS	64	557	8.70	13
41	North Carolina State U	USA	64	821	12.83	14
42	Michigan State U	USA	63	550	8.73	12
43	U Houston	USA	62	404	6.52	9
44	Oklahoma St U Stillwater	USA	62	611	9.85	14
45	Washington St U	USA	61	1,173	19.23	18
46	U Calgary	CAN	59	1,883	31.92	26
47	Leeds Metropolitan U	UK	59	403	6.83	11
48	U South Australia	AUS	56	436	7.79	11
49	La Trobe U	AUS	51	819	16.06	17
50	U Alicante	SPN	50	378	7.56	11

Abbreviations: R = ranking; C = country; TP = total number of publications; TC = total number of citations; TC/TP = citations divided by publications; H = h-index; >100, 50 and 20 = number of papers with more than 100, 50 and 20 citations.

Focusing on productivity through an analysis of the number of papers, we find that the first 5 institutions published 26.5% of papers. Hong Kong Polytechnic University clearly stands out (10%), followed by Texas A&M University (4.7%) and Penn State University (4.3%).

According to the number of citations, the top two institutions remain the same (7.7% and 7.5% respectively). In addition, Texas A&M University shows the highest number of papers in the T100, T50 and T20 citations. Respect to the ratio 'citations per publications, the University of Calgary and the U.S. Forest Service lead the rankings. The institution with the highest h-index is Texas A&M University (39), followed by Hong Kong Polytechnic University and the University of Waterloo with 36, and Penn State University with 32.

4.2 Dynamic Analysis

Table 3 presents a dynamic analysis of the 25 most productive institutions in tourism, leisure and hospitality.

Table 3. Dynamic analysis of the 25 most productive institutions in tourism, leisure and hospitality research

		2000–2004				2005–2009				2010–2014			
R	Institution	TP	TC	TC/TP	H	TP	TC	TC/TP	H	TP	TC	TC/TP	H
1	Hong Kong Pol U	29	1,077	37.14	20	113	2,448	21.66	26	401	2,157	5.38	19
2	Texas A M U Coll	29	1,386	47.79	21	65	1,474	22.68	21	91	628	6.90	12
3	Penn St U	28	974	34.79	16	59	1,006	17.05	19	133	493	3.71	11
4	Griffith U	20	767	38.35	17	37	720	19.46	17	147	774	5.27	14
5	U Waterloo	22	744	33.82	15	64	1,165	18.20	21	59	529	8.97	12
6	Purdue U	6	352	58.67	6	52	1,109	21.33	19	123	488	3.97	11
7	U North Carolina	23	553	24.04	12	40	632	15.80	15	76	196	2.58	7
8	U Queensland	16	505	31.56	14	47	768	16.34	15	108	498	4.61	11
9	U Surrey	17	855	50.29	17	34	735	21.62	15	76	646	8.50	14
10	U Illinois U Ch	28	1,084	38.71	20	45	689	15.31	17	41	122	2.98	6
11	Virginia Poly Inst	14	553	39.50	13	28	632	22.57	10	76	381	5.01	10
12	Kyung Hee U	5	198	39.60	5	25	649	25.96	13	110	455	4.14	9
13	Temple U	–	–	–	–	35	684	19.54	16	90	435	4.83	11
14	Bournemouth U	6	121	20.17	6	26	591	22.73	12	84	430	5.12	10
15	U Central Florida	6	241	40.17	5	31	496	16.00	13	73	314	4.30	10
16	Sejong U	16	269	16.81	9	35	860	24.57	17	64	265	4.14	9
17	Cornell U	–	–	–	–	36	301	8.36	10	72	273	3.79	9
18	Arizona St U	11	675	61.36	9	28	564	20.14	15	35	196	5.60	9
19	U Nevada Las Vegas	6	238	39.67	5	33	436	13.21	10	56	261	4.66	9
20	Clemson U	15	455	30.33	11	18	322	17.89	12	37	120	3.24	7
21	James Cook U	3	114	38.00	3	13	193	14.85	9	53	273	5.15	9
22	U Waikato	9	415	46.11	9	24	364	15.17	12	53	315	5.94	11
23	Monash U	3	207	69.00	3	21	402	19.14	12	59	397	6.73	13
24	Southern Cross U	2	101	50.50	2	16	130	8.13	8	57	272	4.77	9
25	U Florida	16	482	30.13	12	8	167	20.88	7	57	189	3.32	7

Abbreviations are available in Table 2.

Regarding the productivity of the institutions, Hong Kong Polytechnic University stands out as the leader during the past 15 years analyzed; its production is 270% higher than that of the second institution during the last five-year period analyzed. Also notable is the improvement experienced by Purdue University and Penn State University during the past 10 years. Regarding citations received, Hong Kong Polytechnic University again leads the ranking for the past 10 years, followed by Texas A&M University College Station. Based on the ratio 'citations per publications', Hong Kong Polytechnic University lost leadership. Texas A&M University ranks among the top 7 institutions in 4 of the 5 five-year periods, improving its rankings gradually. The h-index also shows the predominance of Hong Kong Polytechnic University and Texas A&M University.

4.3 Bibliographic Coupling and Co-authorship of the Leading Institutions

The leading institutions in tourism, leisure and hospitality research are seen in Figs. 1 and 2.

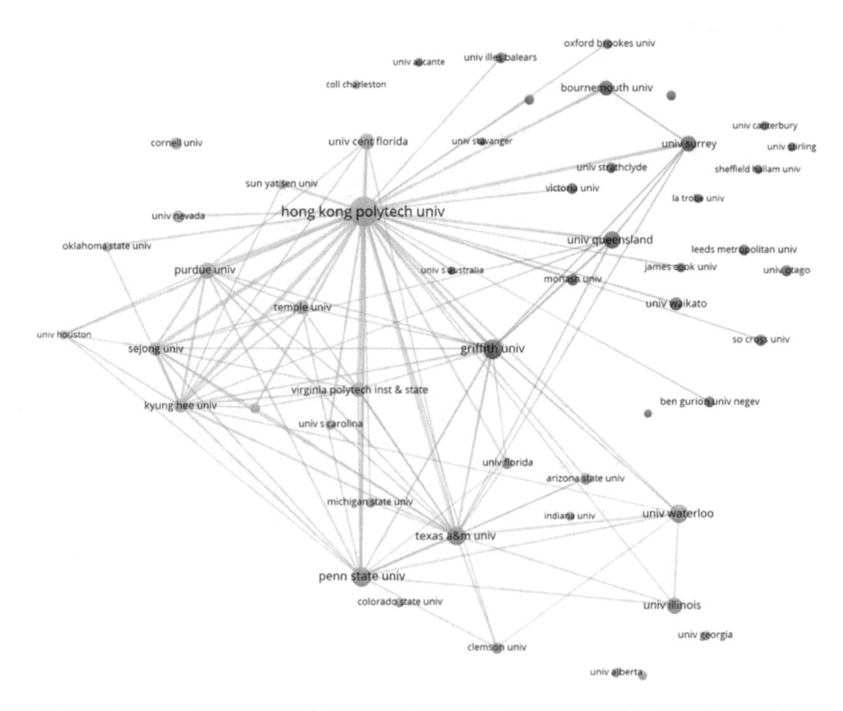

Fig. 1. Mapping of institutions with more than 50 documents and the 100 most influential bibliographic coupling connections. Figure done using Vos viewer software.

There are 3 main nodes with an important interconnection between them. The left node is led by Hong Kong Polytechnic University, the most influential institution of all those analyzed. The majority of institutions in this node are based in the United States and Asia. The upper right node is primarily composed of institutions with European and Australian headquarters, where Griffith University stands out from the rest of institutions. In the lower node we observe a large number of North American institutions; however, none of them stand out over the others.

Next, we perform an analysis of connections via the co-authorship of the institutions analyzed in this paper.

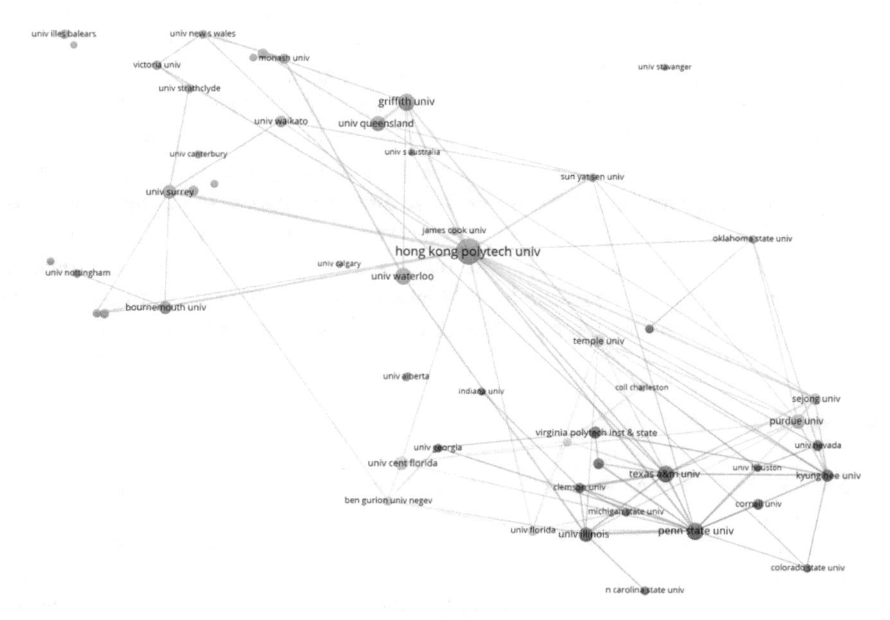

Fig. 2. Mapping of institutions with more than 30 documents and the 100 most influential co-authorship connections. Figure done using Vos viewer software.

Figure 2 shows a map of nodes more dispersed than the previous figure. We can see 6 clusters. Hong Kong Polytechnic University leads, with networks that extend to the rest of the nodes. In certain nodes, the regional connection is more evident and intense than in others, which denotes a stronger relation among authors who belong to nearby geographical areas.

5 Conclusions

We conducted a bibliometric study of the most prolific and influential institutions in tourism, leisure and hospitality, considering the main journals in these fields. Publications in journals focused on tourism, leisure and hospitality have increased exponentially over the past 40 years (Cheng et al. 2011, McKercher and Tung 2015).

The most productive institution was Hong Kong Polytechnic University, although the strong influence of US institutions is also noted. The remainder of the results show a general image of the positions occupied by the main institutions. The dynamic analysis suggests that Hong Kong Polytechnic University has been the most influential institution during the majority of the five-year periods analyzed. Texas A&M University was obtained other important results during the first two five-year periods analyzed, both in number of publications and citations. The bibliometric maps confirm that Hong Kong Polytechnic University is the most influential institution, maintaining important connections with the other institutions. The remaining institutions, with some exceptions, show a greater relation among authors in close geographical areas.

Some of the limitations of our paper are similar to those found in other papers. Although this paper has expanded the number of journals analyzed compared to previous studies, there are other journals that have not been considered. Another limitation is that this study only considered journals indexed in WoS. The paper also does not contemplate the totality of the contributions made by the institutions, because it only analyzes articles, reviews, notes and letters. In addition, given the interdisciplinary nature of tourism, several institutions may have published papers in journals outside of the analyzed area, such as psychology and marketing (Jogaratnam et al. 2005).

Despite these limitations, the paper provides a starting point for future bibliometric studies in these fields. Future research could include bibliometric studies focused on the identification of the main authors, countries and published documents in these fields.

References

Benckendorff, P., Zehrer, A.: A network analysis of tourism research. Ann. Tour. Res. **43**, 121–149 (2013)

Cheng, C., Li, X., Petrick, J.F., O'Leary, J.T.: An examination of tourism journal development. Tour. Manag. **32**, 53–61 (2011)

Cheng, M.: Sharing economy: a review and agenda for future research. Int. J. Hosp. Manag. **57**, 60–70 (2016)

Danielsen, A., Delorme Jr., C.D.: Some empirical evidence on the variables associated with the ranking of economics journals. South. Econ. J. **43**(2), 1149–1160 (1976)

Ding, Y., Rousseau, R., Wolfram, D.: Measuring Scholarly Impact: Methods and Practice. Springer, Cham (2014)

Figueroa-Domecq, C., Pritchard, A., Segovia-Pérez, M.: Tourism gender research: a critical accounting. Ann. Tour. Res. **52**, 87–103 (2015)

García-Lillo, F., Úbeda-García, M., Marco-Lajara, B.: The intellectual structure of research in hospitality management: a literature review using bibliometric methods of the journal international journal of hospitality management. Int. J. Hosp. Manag. **52**, 121–130 (2016)

Goodall, A.H.: Highly cited leaders and the performance of research universities. Res. Policy **38** (7), 1079–1092 (2009)

Hirsch, J.E.: An index to quantify an individual's scientific research output. Proc. Nat. Acad. Sci. U.S.A. **46**(102), 16569–16572 (2005)

Jogaratnam, G., Chon, K., McCleary, K., Mena, M., Yoo, J.: An analysis of institutional contributors to three major academic tourism journals: 1992–2001. Tour. Manag. **26**(5), 641–648 (2005a)

Jogaratnam, G., McCleary, K.W., Mena, M.M., Yoo, J.: An analysis of hospitality and tourism research: institutional contributions. J. Hosp. Tour. Res. **29**(3), 356–371 (2005b)

Kim, J., McMillan, S.J.: Evaluation of internet advertising research: a bibliometric analysis of citations from key sources. J. Advert. **37**(1), 99–112 (2008)

Köseoglu, M.A., Sehitoglu, Y., Craft, J.: Academic foundations of hospitality management research with an emerging country focus: a citation and co-citation analysis. Int. J. Hosp. Manag. **45**, 130–144 (2015a)

Köseoglu, M.A., Sehitoglu, Y., Parnell, J.A.: A bibliometric analysis of scholarly work in leading tourism and hospitality journals: the case of Turkey. Anatolia: Int. J. Tour. Hosp. Res. **26**(3), 359–371 (2015b)

Laengle, S., Merigó, J.M., Miranda, J., Słowínski, R., Bomze, I., Borgonovo, E., Dyson, R., Oliveira, J.F., Teunter, R.: Forty years of the European journal of operational research: a bibliometric overview. Eur. J. Oper. Res. **262**, 803–816 (2017)

Leung, X.Y., Sun, J., Bai, B.: Bibliometrics of social media research: a co-citation and co-wordanalysis. Int. J. Hosp. Manag. **66**, 35–45 (2017)

Mason, D.D.M., Cameron, A.: An analysis of refereed articles in hospitality and the role of editorial members. J. Hosp. Tour. Educ. **18**(1), 11–18 (2006)

McKercher, B.: A case for ranking tourism journals. Tour. Manag. **26**(5), 649–651 (2005)

McKercher, B.: A citation analysis of tourism scholars. Tour. Manag. **29**(6), 1226–1232 (2008)

McKercher, B., Tung, V.: Publishing in tourism and hospitality journals: is the past a prelude to the future? Tour. Manag. **50**, 306–315 (2015)

Merigó, J.M., Yang, J.B.: Accounting research: a bibliometric analysis. Aust. Account. Rev. **80** (27), 71–100 (2017)

Merigó, J.M., Blanco-Mesa, F., Gil-Lafuente, A.M., Yager, R.R.: Thirty years of the international journal of intelligent systems: a bibliometric review. Int. J. Intell. Syst. **32**, 526–554 (2017a)

Merigó, J.M., Gil-Lafuente, A.M., Yager, R.R.: An overview of fuzzy research with bibliometrics indicators. Appl. Soft Comput. **27**, 420–433 (2015a)

Merigó, J.M., Linares-Mustarós, S., Ferrer-Comalat, J.C.: Guest editorial. Kybernetes **46**(1), 2–7 (2017b)

Merigó, J.M., Mas-Tur, A., Roig-Tierno, N., Ribeiro-Soriano, D.: A bibliometric overview of the journal of business research between 1973 and 2014. J. Bus. Res. **68**(12), 2645–2653 (2015b)

Omerzel, D.: A systematic review of research on innovation in hospitality and tourism. Int. J. Contemp. Hosp. Manag. **28**(3), 516–558 (2016)

Podsakoff, P.M., MacKenzie, S.B., Podsakoff, N.P., Bachrach, D.G.: Scholarly influence in the field of management: a bibliometric analysis of the determinants of university and author impact in the management literature in the past quarter century. J. Manag. **34**(4), 641–720 (2008)

Pritchard, A.: Statistical bibliography or bibliometrics. J. Doc. **25**(4), 348–349 (1969)

Ruhanen, L., Weiler, B., Moyle, B.D., McLennan, C.J.: Trends and patterns in sustainable tourism research: a 25-year bibliometric analysis. J. Sustain. Tour. **23**(4), 517–535 (2015)

Samiee, S., Chabowski, B.: Knowledge structure in international marketing: a multi-method bibliometric analysis. J. Acad. Mark. Sci. **40**(2), 364–386 (2012)

Shao-jie, Z., Peng-hui, L., Yan, Y.: Global geographical and scientometric analysis of tourism-themed research. Scientometrics **1**(105), 385–401 (2015)

Valenzuela, L.M., Merigó, J.M., Johnson, W.J., Nicolas, C., Jaramillo, J.F.: Thirty years of the journal of business & industrial marketing: a bibliometric analysis. J. Bus. Industr. Mark. **32** (1), 1–18 (2017)

Van Eck, N.J., Waltman, L.: Software survey: VOS viewer, a computer program for bibliometric mapping. Scientometrics **84**(2), 523–538 (2010)

Vogel, R., Güttel, W.H.: The dynamic capability view in strategic management: a bibliometric review. Int. J. Manag. Rev. **15**(4), 426–446 (2013)

Wagstaff, A., Culyer, A.J.: Four decades of health economics through a bibliometric lens. J. Health Econ. **31**(2), 406–439 (2012)

Ye, Q., Song, H., Li, T.: Cross-institutional collaboration networks in tourism and hospitality research. Tour. Manag. Perspect. **2**(3), 55–64 (2012)

Yuan, Y., Gretzel, U., Tseng, Y.: Revealing the nature of contemporary tourism research: extracting common subject areas through bibliographic coupling. Int. J. Tour. Res. **17**(5), 417–431 (2015)

Yuan, Y., Tseng, Y., Ho, C.: Knowledge base and flow of major tourism and hospitality journals, global review of research in tourism, hospitality and leisure management. Online Int. Res. J. **2** (1), 354–372 (2016)

Zhong, S., Geng, Y., Liu, W., Gao, C., Chen, W.: A bibliometric review on natural resource accounting during 1995–2014. J. Clean. Prod. **139**, 122–132 (2016)

Methods to Analyze Eco-innovation Implementation: A Theoretical Review

Juan F. Pérez-Pérez[1], Jakeline Serrano-García[1](✉), and Juan J. Arbeláez-Toro[2]

[1] Faculty of Economic and Administrative Sciences, Instituto Tecnológico Metropolitano, Medellín, Colombia
juanperez119846@correo.itm.edu.co,
jakelineserrano@itm.edu.co
[2] Faculty of Engineering, Instituto Tecnológico Metropolitano, Medellín, Colombia
juanarbelaez@itm.edu.co

Abstract. Studies and measures to combat the effects of climate change are drawing more and more attention from governments, businesses, and society. The objective of this study is to select a method that allows to evaluate environmentally sustainable strategies that improve production processes. For that purpose, a literature review enabled to identify, in the publications, different methods adopted to address the concept of eco-innovation from different theoretical perspectives. This search revealed methods such as eco-design, multivariate analysis, scenarios, and simulation, which are important to evaluate sustainable innovations and can be applied depending on the objective of each study. Finally, System Dynamics was selected to evaluate policies because it enables to define and design models based on an abstraction of reality, which allows to visualize the possible effects of adopting a new strategy or new organizational structure on an existing system.

Keywords: Eco-innovation · Systems dynamics · Sustainability · Method · Strategies

1 Introduction

In recent decades, a considerable number of summits, conferences, and events have featured sustainable development as their main topic: the United Nations Conference on the Human Environment in Stockholm (1972), the Report of the World Commission on Environment and Development presented in London (WCED 1987), Rio de Janeiro Earth Summit (1992), Kyoto Protocol on climate change (1997), the World Summit on Sustainable Development in Johannesburg (2002), and the Report of the United Nations Environment Programme (PNUMA 2011). The main question in these debates has been how to preserve the environment and natural resources to guarantee them for future generations. As a result, several developed countries (e.g. Norway, the Netherlands, and Germany) and developing nations (such as Qatar) have implemented a series of policies aimed at boosting sustainable economic growth. However, such policies often fail to include in their scope regulations on resource depletion,

J. C. Ferrer-Comalat et al. (Eds.): MS-18 2018, AISC 894, pp. 153–168, 2020.
https://doi.org/10.1007/978-3-030-15413-4_12

environmental pollution, and social injustice. Nevertheless, such interest in sustainable growth has been the driving force behind a rising number of scientific publications that study the relationship between the production of goods, services, and the environment (Beltrán-Esteve and Picazo-Tadeo 2015). Similarly, the scientific community seems to agree on the idea that these changes are unmistakably caused by an increase in the concentrations of greenhouse gases (GHG), mainly as a consequence of different human activities (Beltrán-Esteve and Picazo-Tadeo 2015).

The conventional idea of economic development is associated with an accumulation of goods and services and it stresses the concept of development based on sustainability, which involves satisfying the needs of the present while guaranteeing that future generations can meet their own needs (Beltrán-Esteve and Picazo-Tadeo 2015; WCED 1987, p. 43). Globalization, an increase in the production of goods and services caused by demographic growth, and other related phenomena entail negative consequences, such as exceeding the limits of mass exploitation of natural resources and climate change. An alternative to such practices is to carry out sustainability activities in the production areas of organizations in different countries, regardless of their developmental stage, in order to reconsider design, manufacturing, and building methods by implementing responsible innovation strategies (Tyl et al. 2013).

Therefore, the term "sustainable development" takes precedence in the public discourse to address issues caused by production practices that are hazardous to the environment. In this line, in recent years the term eco-innovation has led to the necessary development of public policies and strategies within organizations to minimize the negative environmental impact that results from production and consumption activities (Jo et al. 2015). According to Rennings (2000), eco-innovation encompasses all the actions that the main stakeholders in the system—such as companies, politicians, and the general community—should conduct to develop new ideas, behaviors, processes, and products, thus contributing to a considerable reduction of the environmental impact and in order to achieve sustainability objectives.

The implementation of eco-innovation processes is an option to mitigate the environmental impact, reduce costs (Arundel and Kemp 2009) and waste, and improve the economic performance of companies. Eco-innovation can also be an option to achieve differentiation among customers with environmental awareness by providing them with products and services that offer better quality and features (Ryszko 2016). Consequently, this type of innovation is the basis to achieve sustainability, especially in the manufacturing industry (Rashid et al. 2015), and it enables companies to directly reduce their carbon footprint (Díaz-García et al. 2015) while raising environmental awareness inside their organizations.

As a result, the main objective of this study is to select a method that enables to evaluate sustainable strategies that improve responsible production and consumption processes. For that purpose, a literature review enabled to analyze different methods, adopted in the past to assess eco-innovations, that facilitate their implementation.

This article is divided as follows: The first part is a review of theoretical conceptualizations on eco-innovation. The second section analyzes methods in the literature that deal with the key concept under study. Afterward, a method is selected because it meets specific characteristics to evaluate future strategies and enables the implementation of eco-innovations in a specific sector. The conclusions and recommendations of this work are found at the end.

2 Theoretical Conceptualization of Eco-innovation

Eco-innovation refers to the goods, services, or processes of an organization that contribute to sustainable development. It means using the available knowledge to modify industrial manufacturing processes with the aim of producing an environmental benefit. American authors Claude Fussler and Peter James coined and defined this concept in the 1990's as the creation of value for the organization and those that benefit from the goods and services it provides, while minimizing the negative risk generated for the environment (Torres-Rivera and Cuevas-Zúñiga 2012).

Similarly, Cheng and Shiu (2012) define eco-innovation as the production, assimilation, or exploitation of products, production processes, services, or managerial or business methods that are novel to an organization (that develops or adopts them) and reduce environmental risks and the negative effects of resource consumption along their life cycle. Authors such as Bleischwitz et al. (2009) maintain that the essential objective of eco-innovation should be the reduction of the flow of materials caused by human activities and the promotion of sustainability objectives.

Likewise, the Organisation for Economic Co-operation and Development (OECD) clarified the meaning of eco-innovation by stressing the two characteristics that set it apart: First, "it is an innovation that reflects an explicit emphasis on the reduction of the environmental impact". Second, "it is not limited to products, processes, or organizational methods; instead, it includes innovations in social and institutional structures" (Organisation for Economic Cooperation and Development 2009).

Three types of eco-innovation can be identified: (1) Eco-friendly processes related to new production methods, e.g. zero CO2 emissions, zero losses, and eco-efficiency in natural resources management; (2) eco-products, i.e. innovations, improvements, or radical changes to existing goods by means of eco-design, sustainable technologies, and reverse engineering to minimize their environmental impact; and (3) organizational eco-innovation, which involves new programs and techniques for organizational systems and tools to evaluate production life cycles, cleaner production, and sustainable consumption (de Oliveira Brasil et al. 2016).

Finally, according to this theoretical framework, eco-innovation refers to the implementation of sustainable processes, procedures, product design, and industrial activities that reduce the negative environmental impact and enable to preserve natural resources.

3 Method

Scopus was used for this literature review because it is a peer-reviewed source of information with one of the highest impact factors; in addition, this database has the largest number of indexed journals (Granda-Orive et al. 2013).

During the first stage, the authors selected terminology related to the main concept under study and keywords that refer to methods or methodologies. The following search equation was the product of that process: TITLE-ABS-KEY (("eco-innovation" OR "sustainable innovation" OR "ecological innovation" OR "green innovation" OR “environmental innovation”) AND (tools OR method* OR model OR technique)). This query retrieved 1,138 results.

Subsequently, in 2017, they were filtered by year of publication since 2004 because the interest in scientific production increased after that period. This step resulted in a list of 184 candidates in Scopus. Regarding subfields, topics that were not related to the concept under study were discarded; in turn, the fields of business, administration, energy, economics, and environmental and decision sciences were included. This study focused on a specific type of publication, scientific articles, in order to narrow down the results. At the end of the process, a total of 488 articles were retrieved.

Afterward, a software application, VOSviewer, was used to build bibliometric networks based on the results obtained from Scopus and show the most-cited publications. Next, the authors read the titles, keywords, and abstracts of said texts (90 in total).

Other restrictions were set and less relevant results were discarded (selection stage). This study includes articles about innovations or technologies focused on sustainability, in any sector, that apply a method to study or analyze the information and the environmental, social, and economic aspects of sustainability. Finally, a total of 35 articles were deemed important to carry out the subsequent analysis.

4 Results

Nine methods were identified among the articles analyzed in the previous stage: TRIZ (Teoriya Resheniya Izobreatatelskikh Zadatch), LCA (Life Cycle Analysis), Factorial Analysis, Regression Analysis, Structural Equation Modeling, Prospective, Multi-Criteria Analysis, System Dynamics, and Agent-Based Modeling. They are used to address the concept of eco-innovation from several perspectives, which are classified in Table 1.

Table 1. Methods identified in the literature. Source: authors’ own work.

Methods to analyze the context of eco-innovation	Ecodesign	TRIZ
		Life Cycle Analysis
	Multivariate analysis	Structural Equation Modeling
		Factorial Analysis
		Regression Analysis
	Scenarios	Multicriteria Analysis
		Prospective
	Simulation	System Dynamics
		Agent-Based Modeling

The following section describes the methods above in order to understand the way they address the implementation of eco-innovations from different theoretical perspectives.

4.1 Ecodesign Methods

Since corporate social responsibility requires more environmentally friendly processes, eco-design is considered fundamental in the manufacturing industry (Kobayashi 2018) to eliminate the possible negative environmental impact (Hur et al. 2005; Russo et al. 2014). For that purpose, a series of methods and tools have been developed to help designers in different organizations create eco-friendly products (Tao et al. 2018), incorporating environmental features into the initial stages of the process.

Among the methods focused on eco-design, companies have adopted LCA as a means to measure the environmental impact of a product, from inception, design, raw material extraction and use, to the final result, which is the disposal of it as waste (Cheng and Shiu 2012; Huber 2008; Umeda et al. 2012). To conduct the LCA of a product, designers should consider not only said cycle but also several important aspects such as the law, customers' needs, strategic objectives of the organization, and market and technology trends (Umeda et al. 2012) without compromising the features of the product (Russo et al. 2014; Tao et al. 2018) as a result of material modifications.

Therefore, an environmental evaluation includes the identification, quantification, evaluation, and prioritization of environmental aspects related to a production system, as well as its conceptual design (Hur et al. 2005; Kobayashi 2018). However, several methods can be adopted to identify environmental aspects and none of them is preferred over the others (Hur et al. 2005). For that reason, LCA has been integrated with other methods, such as TRIZ (Kobayashi 2018; Russo et al. 2014), CAD (Computer-Aided Design), CAE (Computer-Aided Engineering), and optimization, which allow to effectively add environmental attributes to the design of the product (Tao et al. 2018).

As previously mentioned, a series of available methods help designers evaluate the life cycle of products or aim at providing suggestions on how to design products (Russo et al. 2011) in accordance with sustainable objectives set by the company or to comply with the applicable regulations in force. TRIZ (acronym of *Theory of Inventive Problem Solving* in Russian) is another remarkable method to design eco-friendly products. This approach is based on universal ways to solve problems and people's capacity to design innovative solutions (Kobayashi 2018; Yang and Chen 2011). It also offers a toolbox for designers to avoid trial and error processes during the design stages and solve problems inventively (Yang and Chen 2011).

In this context, Russo et al. (2011) introduced a way to use TRIZ to evaluate and innovate in a technical system in order to integrate daily design activities and achieve sustainable results. In the same line, Yang and Chen (2012) proposed to solve design problems using TRIZ tools and generated new ideas by integrating them with Case-Based Reasoning, since the latter aims at imitating the human capacity to remember past experiences that allow to solve problems. Russo et al. (2014) designed a tool called "iTree" combining instructions derived from problem solving, conceptual design approaches, better sustainable experiences, and simulation tools such as CAD.

In general, eco-innovation activities can be divided into two phases: (1) evaluation of the product the company already has and its environmental impact and (2) improvement of the product incorporating some design considerations. The main objective of a LCA is to considerably reduce energy and resource consumption (Tao et al. 2018; Umeda et al. 2012), emissions, and the use of hazardous substances, but authors such as Russo et al. (2015) mention its complexity. Additionally, according to the literature, TRIZ and LCA have been tested by experts and shown excellent results in the design of eco-friendly products, depending on their approach and possible combination with other methods.

4.2 Multi-variate Analysis Methods

Environmental innovation has drawn considerable attention from all sectors. Furthermore, it has been the force behind a considerable amount of empirical and theoretical literature with quantitative and qualitative approaches (Cuerva et al. 2014), which has been necessary to find statistically valid relationships and improve decision making. For instance, an empirical analysis by Pujari (2006) that examined the results of a survey on projects to develop new products in the United States enabled to explore the relationship and the effects of eco-innovation activities on the performance of new eco-friendly products. The data in that empirical study were analyzed by means of hierarchical regression to find statistically significant relationships between market performance and several independent factors.

Cuerva et al. (2014) made another contribution in the same line. They examined differences between the factors that influence the adoption of "green" and "non-green" innovations by means of a bivariant probit regression model, which allows to estimate the probability of two different but correlated processes. In turn, Sezen and Çankaya (2013) delved into the influence of improving organizational processes and eco-innovation on the sustainable performance of a firm. They suggested hypotheses based on data generated by 53 Turkish companies in different sectors. Afterward, a regression analysis was carried out to statistically prove the hypotheses and define the direction of the relationships established by said authors.

Environmental innovations include all kinds of economic, technological, legal, organizational, and behavioral advances that enable to reduce environmental damage (Cuerva et al. 2014). Nevertheless, it is necessary to evaluate possible relationships between individual factors related to the behavior and the will to innovate of an organization in order to achieve greener goals. For that purpose, statistical validations offer possibilities of analysis according to the objective of each study.

Factorial Analysis is another multi-variate method to validate data that has been employed to evaluate eco-innovations in the industrial sector. In this respect, Cheng and Shiu (2012) provided an instrument focused on the implementation of eco-innovation, which was understood as a series of actions that aim at generating specific projects that help to mitigate the environmental damage produced by industrial activities. They used factorial analysis to identify the dimensionality of eco-innovation and correlations among variables defined in the study.

Similarly, Cheng et al. (2014) studied the interrelationships among three types of eco-innovation (process, product, and organization) and the impact they may have on organizational performance. They employed structural equation models with data collected from 121 companies in Taiwan to reveal the direct, indirect, and total effects of said types of eco-innovation. Likewise, de Oliveira Brasil et al. (2016) examined, also by means of structural equation models, the possible relationships among those three kinds of innovation. In that case, the study was empirically conducted at 70 textile companies in Brazil and its results show the direct effects of organizational and product eco-innovations on a company's performance. Furthermore, the indirect effects of these processes on commercial performance can also be observed and the advantages of each type of innovation are highlighted. This leads to the conclusion that companies should develop organizational eco-innovations as a first step to develop the necessary infrastructure and knowledge to foster the other types of eco-friendly innovations (Cheng et al. 2014; de Oliveira Brasil et al. 2016).

Furthermore, based on an empirical study of the Brazilian agroindustrial sector, Cunico et al. (2017) demonstrated that technological cooperation generates eco-innovations by analyzing the degree of participation of each stakeholder (universities, companies, and the government) in said interactions. By applying a model of structural equations, they positively confirmed the participation of the stakeholders in technological cooperation for the generation of eco-innovations, thus demonstrating the degree of influence of the independent variables on the dependent counterparts in the model.

4.2.1 Scenario Building Method

Current industrial ecology research mainly focuses its efforts on measuring the impact of implementing eco-friendly practices. For that reason, Allwood et al. (2008) proposed a different approach that considers possible practical changes in production (such as new product design) by means of what they call a *catalog*, i.e. a series of scenarios that help sustainability. They expressed great interest in using scenario analysis as a means to visualize plausible future paths of sustainable development because they have recently widened. The paradigm of sustainable development inherently encompasses research and future thinking, because the definitions and goals refer to both present and future generations. According to Nieto-Romero et al. (2016), scenario planning has become an important tool to combat the design of uncertain futures towards sustainability.

Likewise, Dias et al. (2016) stress the importance of sustainability in industrial activities because, as can be observed throughout history, resources are being depleted. Because of this outlook, future plans should be changed to affirm our commitment with the environment and the conservation of resources for future generations. The main objective of their study is to analyze methods to produce images and enlarge their frameworks (future images) including sustainable dimensions. Their method was interdisciplinary work that connects concepts from different fields. This conceptual basis enables to determine if it will be possible to propose a sustainable alternative to the methods employed to build images. Afterward, they present a study case to test the new scenario frameworks that will be applied to the Brazilian biodiesel industry with a time horizon until 2030. As a result of their work, 4 scenarios were built with the support of expert staff and political decision makers. Such scenarios include options that range from sector sustainability and social inclusion to actions that would mean the collapse of civilization.

Another approach in the same line was described by Saritas and Aylen (2010). They combined scenarios and roadmapping to develop cleaner production standards for metal manufacturing based on the European project CLEANPROD. The objective of the project was to develop a series of roadmaps to achieve sustainability in machining, coating, and surface preparation at metal manufacturing organizations. Their approach proposes the use of scenarios and roadmapping by means of: (1) establishing visions for the roadmaps considering alternative futures, which adds an exploratory feature to the process; (2) alternative paths for those roadmaps; (3) improving the explanatory power of roadmaps, and (4) testing the robustness of the roadmapping compared to the scenarios.

Both methods have advantages and disadvantages, which highlights the need to integrate them and present a new approach. According to Saritas and Aylen (2010):

> The integration of scenarios and roadmapping is beneficial in a policy and strategy making process. Both methods have desirable properties and are complementary. In this respect, the use of scenarios in the roadmapping process helps overcome some of the criticisms directed against roadmapping as a foresight method. With the introduction of scenarios, the roadmapping process is not only normative, but also becomes exploratory by considering a set of probable futures. The linearity and isolation of roadmaps are eliminated with the application of a creative, interactive and collaborative scenario planning process (p. 1067).

To conduct a sustainability analysis of an organization or society at large, a series of indicators must be established to be able to trace decisions made on a specific regard. Such indicators should include economic, environmental, and social dimensions (Scarpellini et al. 2013; Stoycheva et al. 2018) that are influenced by the decisions of stakeholders. As an alternative to integrate the main stakeholders in the areas mentioned before, multi-criteria decision making or analyses are employed. The latter have become more popular in sustainable management (Scarpellini et al. 2013) because they enable to consider a great number of data, relationships, objects, and the preferences of the parties interested in the situation under analysis (Stoycheva et al. 2018). Multi-criteria analyses quantify the dimensions, which can be used to compare alternatives. This type of comparison is important when sustainable innovations are analyzed because sustainability criteria are complex to quantify (Stoycheva et al. 2018).

Subsequently, Scarpellini et al. (2013) presented a simplified multi-criteria decision analysis based on savings-investment-employment curves to analyze sectors that are open to sustainability in economic, social, and environmental terms and thus identify the way actions are prioritized. Similarly, Stoycheva et al. (2018) presented a conceptual framework that enabled to evaluate eco-friendly manufacturing and its application in the automotive sector in order to select cleaner manufacturing alternatives. The results included alternative materials for production, which can be quantitatively selected according to the organization's sustainable objectives. Watróbski (2016) drafted a process to select a subset of multi-criteria analysis methods adapted to particular problems of green logistics. To conduct that study, its author selected a series of logistic problems that were addressed by means of a literature review validated with the knowledge of experts.

4.2.2 Modeling and Simulation Methods

Recent years have seen growing environmental problems caused by pollution and, as a result, strategies should be promoted to reduce contamination adopting efficient methods (Tian et al. 2014). Prospective and simulation methods, as well as variations, have been adopted to build scenarios to recreate the ideal conditions of systems. One such method is Agent-Based Modeling (ABM), which thoroughly examines the behavior of an agent and a set of data that enable to model complex systems (Rai and Robinson 2015).

Rai and Robinson (2015) designed an agent-based technology adoption model that was mainly applied to the adoption of residential photovoltaic systems in Austin, Texas, between 2004 and 2013. The objective was to empirically generate temporal and spatial patterns as basis to improve decision making regarding public utilities planning. Schwarz and Ernst (2009) presented an ABM of the dissemination of three water-saving innovations in Southern Germany. Data resulting from an empirical questionnaire with 272 participants were used to create the model. Such study interpreted households as agents that make decisions related to the use of water-saving technologies. For the model, however, an agent does not represent a specific household, but a group of households with particular shared characteristics within a square mile. The result of the study was the design of four possible scenarios to disseminate this type of technologies.

Likewise, Desmarchelier et al. (2013) proposed a theoretical ABM that examined the extent to which environmental taxation and end-user information policies can promote eco-innovation in the service sector. Their simulations revealed that environmental policies motivate companies in the service sector to generate sustainable innovations in order to offer them to their users and improve their processes.

Other studies have employed System Dynamics (SD) to simulate and verify the effectiveness of policies in different social sectors. One of them is the energy policy of the United States, which was analyzed by Naill (1992). Said author constructed an integrated model of energy supply and demand in that country. According to this expert, this topic of study (energy consumption in the United States) has been and will be of great interest due to the high levels of pollution it generates and the import of vast amounts of petroleum.

Furthermore, Kazemi and Hosseinzadeh (2016) proposed to plan energy provision in Iran by using an SD model. Their process generated forecasts of future GHG emissions trends after the implementation of a series of combined policies that promote the reduction of such gases. Possible future scenarios were analyzed to answer questions posed during the design of the SD model, e.g. "*What if…?*". In turn, the answers led to a list of policies related to GHG reduction. Therefore, this model is proposed as a tool that enables policy makers to estimate the consequences of Iranian energy consumption and reduce GHG in the long term.

Also using SD, other authors have developed generic models of the dissemination of eco-friendly technologies that can be adopted by any organization (Müller et al. 2013, and Timma et al. 2015). Shih and Tseng (2014) designed a DS model to calculate the amount of energy saved; they used a scenario that considered the promotion of a sustainable energy policy, renewable energies, and improvements in energy efficiency. Besides, the development and verification of said model quantified the benefits from 2010 to 2030. Ansari and Seifi (2013) analyzed the energy consumption and CO2 emissions of the Iranian cement industry by designing another DS model and creating scenarios that included different production and export activities.

According to Shih and Tseng (2014), a series of DS energy models have been adopted to evaluate scenarios and policies regarding CO_2 reduction, as well as their economic consequences. They highlight:

The SD approach is suitable for modeling dynamic environments, such as ecosystems and human activities, on a muti-dimensional scale with time-dependent variables. SD modeling has been applied for strategic energy planning and policy analysis since the early 1970s, starting with the well-known "Limits to Growth" (Shih and Tseng 2014, p. 58).

4.3 Selecting the Method to Analyze Strategies

This literature review confirms that several methods have been employed to address the main topic in this study and, therefore, the selection of an approach depends on the type of problem to be examined. Furthermore, the authors conclude that SD is a good option to model very complex problems. Nevertheless, regarding the analysis of strategies to implement eco-innovation processes (the purpose of selecting a method in this work), SD is evidently different from other analysis methods. This is because SD involves identifying relationships of influence among variables in a complex system in order to compare reality with the dynamic behavior of a model that symbolizes the entire system. Moreover, said method is suitable to simulate systems that include nonlinear relationships generated in dynamic environments and, as a result, employing an analytical approach or method to solve said nonlinearities is infeasible (Poles 2013). Therefore, DS is considered an adequate method to evaluate strategies because it enables to consider variables, dynamic processes, policies, and scenarios that focus on improving organizational performance.

This review of different methods that have been established to analyze sustainable strategies at organizations shows that scenario analysis has enjoyed widespread acceptance and usage as well. However, it is highly complex because scenarios need to be validated by experts (Vergara et al. 2010) and contradictions among them may arise (Godet 2000). On the other hand, SD proves to be more suitable because the same results can be immediately simulated and generated with lower risks and costs since there is no need to gather experts to validate them.

The multi-variate methods described in this work to study the concept of eco-innovation enable researchers to interpret and visualize a set of data, generally big, to find relationships among variables or individuals. In general, the character of these techniques tends to be exploratory and not inferential (Nieto 2015). These methods, as well as SD, mainly focus on analyzing variables and the relationships among them. In some cases, however, very big samples are necessary to reduce the error (Manzano 2000). Some methods, such as regression analysis, demand linearity among variables (Aggarwal and Ranganathan 2017; Sun and Park 2017); besides, the observations of a sample should be independent and not include extreme values because this may generate errors and results may be deceiving (Aggarwal and Ranganathan 2017). Conversely, SD employs extreme conditions to validate and verify the structure of the models.

The objective of ABM is to study the basic components of the system to be analyzed. Consequently, it applies a process of reality conceptualization to each element individually. In turn, SD focuses on the entire system, analyzing the relationships among observable variables (Izquierdo et al. 2008). In general, one may say that SD is used to model homogeneous entities and ABM can analyze complex systems that require the interaction of heterogeneous individuals who need to exchange information.

On the other hand, the design methods that address the concept of eco-innovation presented in this work fail to achieve the objective of this study because they are employed to design eco-friendly products and not to evaluate organizational strategies. Table 2 shows the description and limitations of the methodologies studied.

SD models are the most suitable option to study systems that involve interactions between physical and human factors (Agarwal et al. 2002). Moreover, these methods are useful to reduce the uncertainty of decision making when the system is unknown, there is confusion, and no direct solution to the problem can be found (Pidd 1999).

SD simulations allow to analyze systems in situations in which conducting physical experiments is complicated, infeasible, risky, or costly because an investment in human, economic, and technology resources is necessary. Simulating said complex systems provides a better understanding of their behavior, and new variables or parameters could be incorporated to improve the system. Furthermore, simulations enable to conduct tests and obtain results quickly and change variables that may generate uncertainty in the researcher or the director of an organization. As a result, scenario simulation can anticipate possible events the organization will experience (Aguirre and Ramírez 2014).

Table 2. Description and limitations of the methods. Source: authors' own work.

Method	Description	Limitations
LCA	Method to quantify the environmental effects of the product throughout its life cycle (Tao et al. 2018)	Weak penetration and usually complex (Russo et al. 2015)
TRIZ	Feasible way to identify a new eco-product design (Yang and Chen 2012)	Still lacks a systemic vision of the product life cycle (Russo et al. 2014)
Regression Analysis	Studies statistically significant relationships (Pujari 2006)	Assumes uniform linearity among dependent and independent variables; therefore, several types of reality cannot be represented (Sun and Park 2017)
Factorial Analysis	Employed to prove theoretically relevant hypotheses and applied to the analysis of data collected by statistical instruments (Nieto 2015)	Requires an ideal sample size between 300 and 400 cases to minimize the probability of errors. Samples under 50 cases should be avoided (Mavrou 2015)
Structural Equations	Establishes measurement and structural models to analyze complex relationships of human behavior (Calvo-Porral et al. 2013)	The sample size should be big ($n > 250$); at least 10 cases should be included for each observed variable (Manzano 2000)

(continued)

Table 2. (*continued*)

Method	Description	Limitations
Prospective	Based on a systemic and holistic perspective that enables to build alternative future scenarios (Rodríguez 2015)	Scenario building has not had a solid theoretical foundation; for that reason, its application has been restricted to small groups of experts with subjective criteria (Vergara et al. 2010)
Multi-Criteria Analysis	Includes a vast number of data, relationships, and objectives as they occur (Scarpellini et al. 2013)	The results depend on the available information and the way it is organized, the selected method, and the preferences of the decision maker (Scarpellini et al. 2013)
System Dynamics	Enables to create models of complex systems characterized by accumulations (delays) and nonlinear feedback mechanisms that explicitly reflect the relationship between cause and effect (Timma et al. 2015; Aizstrauta et al. 2015)	The analysis is not focused on the behavior of individuals but on that of a system. This method is not used to analyze homogeneous entities, and it does not include space as an essential variable (Díez-Echavarría 2015)
Agent-Based Modeling	Studies the evolution of complex human-technical systems due to the flexibility it offers to describe in detail the behavioral and structural aspects (policy, prices, and infrastructure) of said systems (Rai and Robinson 2015)	When the model contains more variables and the agents have many behavior rules, it is more complex to find a way to validate the results (Aguirre and Ramírez 2014)

5 Conclusions

Eco-innovation has characteristics different from other types of development because it constitutes a requirement for long-term industrial growth. In that regard, eco-innovation refers to the goods, services, or processes of an organization that contribute to sustainable development.

Companies consider environmental regulations a threat to their competitive capacity due to the costs associated with compliance. However, there is a growing body of evidence and arguments that suggests that eco-innovation can mitigate the traditional confrontation between competitiveness and environmental protection because it tends to improve efficiency and cost management, open new markets, and reduce the environmental impact.

This state-of-the-art review analyzed eco-innovation from the perspective of several methods in different fields. Said methods offer application alternatives that range from designing eco-friendly products (that help to minimize the environmental impact of daily-consumed products) to empirical analyses that present organizational circumstances resulting from the implementation of sustainable processes. Thus, a series of scenarios are generated and simulated adopting different methods that improve decision making in order to provide a general outlook of these issues.

The objective of this review was to identify a method that enabled to evaluate implementation strategies of eco-innovation processes. Based on the results above, the SD method was found to exhibit all the necessary features to estimate possible scenarios because it allows to simulate complex systems and consider delays, which are to be expected when a measure or strategy is implemented.

The SD method is different from other modeling techniques because its application does not require a predetermined model; instead, the model is defined and designed based on an abstraction of reality that varies with each phenomenon or problem under study. In summary, SD-based simulation enables to forecast the possible effects of adopting a new strategy or organizational structure on a real system.

References

Agarwal, C., Green, G.M., Grove, J.M., Evans, T.P., Schweik, C.M.: A review and assessment of land-use change models: dynamics of space, time, and human choice. Apollo Int. Mag. Art Antiq. **1**, 62 (2002). https://doi.org/10.1289/ehp.6514

Aggarwal, R., Ranganathan, P.: Common pitfalls in statistical analysis: linear regression analysis. Perspect. Clin. Res. **8**(2), 100 (2017). https://doi.org/10.4103/2229-3485.203040

Aguirre, J., Ramírez, M.: Análisis cienciométrico de modelación y simulación de Sistemas de innovación. Instituto Tecnológico Metropolitano, Medellín (2014). https://doi.org/10.22430/9789588743615

Aizstrauta, D., Ginters, E., Eroles, M.A.P.: Applying theory of diffusion of innovations to evaluate technology acceptance and sustainability. Procedia Comput. Sci. **43**(C), 69–77 (2015). https://doi.org/10.1016/j.procs.2014.12.010

Allwood, J., Laursen, S., Russell, S., de Rodríguez, C.M., Bocken, N.M.: An approach to scenario analysis of the sustainability of an industrial sector applied to clothing and textiles in the UK. J. Clean. Prod. **16**(12), 1234–1246 (2008). https://doi.org/10.1016/j.jclepro.2007.06.014

Ansari, N., Seifi, A.: A system dynamics model for analyzing energy consumption and CO_2 emission in Iranian cement industry under various production and export scenarios. Energy Policy **58**, 75–89 (2013). https://doi.org/10.1016/j.enpol.2013.02.042

Arundel, A., Kemp, R.: Measuring eco-innovation. Evaluation **49**(89), 1–40 (2009)

Beltrán-Esteve, M., Picazo-Tadeo, A.J.: Assessing environmental performance trends in the transport industry: eco-innovation or catching-up? Energy Econ. **51**, 570–580 (2015). https://doi.org/10.1016/j.eneco.2015.08.018

Bleischwitz, R., Giljum, S., Kuhndt, M., Schmidt-Bleek, F.: Eco-innovation – putting the EU on the path to a resource and energy efficient economy (2009). seri.at/wp-content/…/06/European-Parliament-2009-EcoInnovation.pdf

Calvo-Porral, C., Martínez-Fernández, V.-A., Juanatey-Boga, O.: Análisis de dos modelos de ecuaciones estructurales alternativos para medir la intención de compra. Revista Investigación Operacional **34**(3), 230–243 (2013). http://rev-inv-ope.univ-paris1.fr/files/34313/34313-05.pdf

Cheng, C.C.J., Yang, C.L., Sheu, C.: The link between eco-innovation and business performance: a Taiwanese industry context. J. Clean. Prod. **64**, 81–90 (2014). https://doi.org/10.1016/j.jclepro.2013.09.050

Cheng, C.C., Shiu, E.C.: Validation of a proposed instrument for measuring eco-innovation: an implementation perspective. Technovation **32**(6), 329–344 (2012). https://doi.org/10.1016/j.technovation.2012.02.001

Cuerva, M.C., Triguero-Cano, Á., Córcoles, D.: Drivers of green and non-green innovation: empirical evidence in low-tech SMEs. J. Clean. Prod. **68**, 104–113 (2014). https://doi.org/10.1016/j.jclepro.2013.10.049

Cunico, E., Cirani, C.B.S., Lopes, E.L., Jabbour, C.J.C.: Eco-innovation and technological cooperation in cassava processing companies: structural equation modeling. Revista de Administração **52**(1), 36–46 (2017). https://doi.org/10.1016/j.rausp.2016.09.006

de Oliveira Brasil, M.V., Sá de Abreu, M.C., Lázaro da Silva Filho, J.C., Leocádio, A.L.: Relationship between eco-innovations and the impact on business. Revista de Administracao **51**, 276–287 (2016). https://doi.org/10.1016/j.rausp.2016.06.003

Desmarchelier, B., Djellal, F., Gallouj, F.: Environmental policies and eco-innovations by service firms: an agent-based model. Technol. Forecast. Soc. Chang. **80**(7), 1395–1408 (2013). https://doi.org/10.1016/j.techfore.2012.11.005

de Paula Dias, M.A., de Souza Vianna, J.N., Felby, C.: Sustainability in the prospective scenarios methods: a case study of scenarios for biodiesel industry in Brazil, for 2030. Futures **82**, 1–14 (2016). https://doi.org/10.1016/j.futures.2016.06.005

Díaz-García, C., González-Moreno, Á., Sáez-Martínez, F.J.: Eco-innovation: insights from a literature review. Innov. Manag. Policy Pract. **17**(1), 6–23 (2015). https://doi.org/10.1080/14479338.2015.1011060

Díez-Echavarría, L.F.: Modeling the relationships between variability and climate change in the incidence of malaria in Colombia: a comparative analysis between system dynamics and agent-based simulation. Universidad Nacional de Colombia (2015)

Godet, M.: La caja de herramientas de la prospectiva estratégica (2000)

Granda-Orive, J.I., Alonso-Arroyo, A., García-Río, F., Solano-Reina, S., Jiménez-Ruiz, C.A., Aleixandre-Benavent, R.: Ciertas ventajas de Scopus sobre Web of Science en un análisis bibliométrico sobre tabaquismo. Revista Española de Documentación Científica **36**(2), 9 (2013). https://doi.org/10.3989/redc.2013.2.941

Huber, J.: Technological environmental innovations (TEIs) in a chain-analytical and life-cycle-analytical perspective. J. Clean. Prod. **16**(18), 1980–1986 (2008). https://doi.org/10.1016/j.jclepro.2008.01.014

Hur, T., Lee, J., Ryu, J., Kwon, E.: Simplified LCA and matrix methods in identifying the environmental aspects of a product system. J. Environ. Manag. **75**(3), 229–237 (2005). https://doi.org/10.1016/j.jenvman.2004.11.014

Izquierdo, L., Galán, J., Santos, J., Del Olmo, R.: Modelado de sistemas complejos mediante simulación basada en agentes y mediante dinámica de sistemas. Red Cientifica, **20**(1), 46–66 (2008). https://doi.org/2017-01-18

Jo, J.H., Roh, T.W., Kim, S., Youn, Y.C., Park, M.S., Han, K.J., Jang, E.K.: Eco-innovation for sustainability: evidence from 49 countries in Asia and Europe. Sustainability (Switzerland) **7** (12), 16820–16835 (2015). https://doi.org/10.3390/su71215849

Kazemi, A., Hosseinzadeh, M.: Policy analysis of greenhouse gases' mitigation in Iran energy sector using system dynamics approach. Environ. Progress Sustain. Energy, **35**(4), 1221–1230 (2016). https://doi.org/10.1002/ep.12355

Kobayashi, H.: A systematic approach to eco-innovative product design based on life cycle planning. Adv. Eng. Inform. **20**(2), 113–125 (2018). https://doi.org/10.1016/j.aei.2005.11.002

Manzano, A.: Introducción a los modelos de ecuaciones estructurales. **7**, 67–72 (2000)

Mavrou, I.: Exploratory factor analysis: conceptual and methodological issues. Revista Nebrija de Lingüística Aplicada a La Enseñanza de Las Lenguas **19** (2015). http://www.nebrija.com/revista-linguistica/analisis-factorial-exploratorio.html

Müller, M.O., Kaufmann-Hayoz, R., Schwaninger, M., Ulli-Beer, S.: The diffusion of eco-technologies: a model-based theory. Underst. Complex Syst. pp. 49–67 (2013). https://doi.org/10.1007/978-1-4614-8606-0-4

Naill, R.F.: System dynamics model for national energy policy planning. Syst. Dyn. Rev. **8**(1), 1–19 (1992). https://doi.org/10.1002/sdr.4260080102

Nieto-Romero, M., Milcu, A., Leventon, J., Mikulcak, F., Fischer, J.: The role of scenarios in fostering collective action for sustainable development: lessons from central Romania. Land Use Policy **50**, 156–168 (2016). https://doi.org/10.1016/j.landusepol.2015.09.013

Nieto, L.: Análisis Multivariado. Diplomado En Estadística Aplicada, pp. 1–35 (2015)

Organisation for Economic Cooperation and Development: Sustainable Manufacturing and Eco-Innovation: Framework, Practices and Measurement. OECD, 38 (2009). https://doi.org/10.1177/0022146512457153

Pidd, M.: Just modeling through: a rough guide to modeling. Interfaces **29**(2), 118–132 (1999). https://doi.org/10.1287/inte.29.2.118

Poles, R.: System dynamics modelling of a production and inventory system for remanufacturing to evaluate system improvement strategies. Int. J. Prod. Econ. **144**(1), 189–199 (2013). https://doi.org/10.1016/j.ijpe.2013.02.003

Pujari, D.: Eco-innovation and new product development: understanding the influences on market performance. Technovation **26**(1), 76–85 (2006). https://doi.org/10.1016/j.technovation.2004.07.006

Rai, V., Robinson, S.A.: Agent-based modeling of energy technology adoption: empirical integration of social, behavioral, economic, and environmental factors. Environ. Model Softw. **70**, 163–177 (2015). https://doi.org/10.1016/j.envsoft.2015.04.014

Rashid, N., Jabar, J., Yahya, S., Shami, S.: Dynamic eco innovation practices: a systematic review of state of the art and future direction for eco innovation study. Asian Soc. Sci. **11**(1), 8–21 (2015). https://doi.org/10.5539/ass.v11n1p8

Rennings, K.: Redefining innovation - Eco-innovation research and the contribution from ecological economics. Ecol. Econ. **32**(2), 319–332 (2000). https://doi.org/10.1016/S0921-8009(99)00112-3

Rodríguez, C.M.: Pensamiento prospectivo: visión sistémica de la construcción del futuro. Análisis **46**(84), 89–104 (2015). https://doi.org/10.15332/s0120-8454.2014.0084.05

Russo, D., Regazzoni, D., Montecchi, T.: Eco-design with TRIZ laws of evolution. Procedia Eng. **9**, 311–322 (2011). https://doi.org/10.1016/j.proeng.2011.03.121

Russo, D., Rizzi, C., Montelisciani, G.: Inventive guidelines for a TRIZ-based eco-design matrix. J. Clean. Prod. **76**, 95–105 (2014). https://doi.org/10.1016/j.jclepro.2014.04.057

Russo, D., Schöfer, M., Bersano, G.: Supporting ECO-innovation in SMEs by TRIZ Eco-guidelines. Procedia Eng. **131**, 831–839 (2015). https://doi.org/10.1016/j.proeng.2015.12.388

Ryszko, A.: Proactive environmental strategy, technological eco-innovation and firm performance-case of Poland. Sustainability (Switzerland) **8**(2), 156 (2016). https://doi.org/10.3390/su8020156

Saritas, O., Aylen, J.: Using scenarios for roadmapping: the case of clean production. Technol. Forecast. Soc. Chang. (2010). https://doi.org/10.1016/j.techfore.2010.03.003

Scarpellini, S., Valero, A., Llera, E., Aranda, A.: Multicriteria analysis for the assessment of energy innovations in the transport sector. Energy **57**, 160–168 (2013). https://doi.org/10.1016/j.energy.2012.12.004

Schwarz, N., Ernst, A.: Agent-based modeling of the diffusion of environmental innovations - an empirical approach. Technol. Forecast. Soc. Chang. **76**(4), 497–511 (2009). https://doi.org/10.1016/j.techfore.2008.03.024

Sezen, B., Çankaya, S.Y.: Effects of green manufacturing and eco-innovation on sustainability performance. Procedia – Soc. Behav. Sci. **99**, 154–163 (2013). https://doi.org/10.1016/j.sbspro.2013.10.481

Shih, Y., Tseng, C.: Cost-benefit analysis of sustainable energy development using life-cycle co-benefits assessment and the system dynamics approach. Appl. Energy **119**, 57–66 (2014). https://doi.org/10.1016/j.apenergy.2013.12.031

Stoycheva, S., Marchese, D., Paul, C., Padoan, S., Juhmani, A., Linkov, I.: Multi-criteria decision analysis framework for sustainable manufacturing in automotive industry. J. Clean. Prod. **187**, 257–272 (2018). https://doi.org/10.1016/j.jclepro.2018.03.133

Sun, E., Park, S.: The relationship between chaebol and firm value using Bayesian network. J. Appl. Bus. Res. **33**(6), 1113–1128 (2017)

Tao, J., Li, L., Yu, S., Li, L., Yu, S.: An innovative eco-design approach based on integration of LCA, CAD\CAE and optimization. J. Clean. Prod. **187** (2018). https://doi.org/10.1016/j.jclepro.2018.03.213

Tian, Y., Govindan, K., Zhu, Q.: A system dynamics model based on evolutionary game theory for green supply chain management diffusion among Chinese manufacturers. J. Clean. Prod. **80**, 96–105 (2014). https://doi.org/10.1016/j.jclepro.2014.05.076

Timma, L., Bariss, U., Blumberga, A., Blumberga, D.: Outlining innovation diffusion processes in households using system dynamics. case study: energy efficiency lighting. Energy Procedia **75**, 2859–2864 (2015). https://doi.org/10.1016/j.egypro.2015.07.574

Torres-Rivera, A.D., Cuevas-Zúñiga, I.Y.: Propuesta de tratamiento contable de las eco-eficiencias. Revista Del Instituto Internacional de Costos, pp. 187–210 (2012)

Tyl, B., Legardeur, J., Millet, D., Vallet, F.: Adaptation of the creativity tool ASIT to support eco-ideation phases. In: Sustainable Intelligent Manufacturing International Conference, Libon, Portugal (2013). https://doi.org/10.1201/b15002-85

Umeda, Y., Takata, S., Kimura, F., Tomiyama, T., Sutherland, J.W., Kara, S., Duflou, J.R.: Toward integrated product and process life cycle planning - an environmental perspective. CIRP Annals – Manuf. Technol. **61**(2), 681–702 (2012). https://doi.org/10.1016/j.cirp.2012.05.004

Vergara, J.C., Fontalvo, T.J., Maza, F.: La planeación por escenarios: Revisión de conceptos y propuestas metodológicas. Prospect **8**(2), 21–29 (2010). http://altekio.es/wordpress/wp-content/uploads/2013/12/Dialnet-LaPlaneacionPorEscenariosRevisionDeConceptosYPropu-3634575.pdf

Watróbski, J.: Outline of multicriteria decision-making in green logistics. Transp. Res. Procedia **16**(March), 537–552 (2016). https://doi.org/10.1016/j.trpro.2016.11.051

WCED: Our common future: report of the world commission on environment and development. United Nations Commission **4**(1), 300 (1987). https://doi.org/10.1080/07488008808408783

Yang, C.J., Chen, J.L.: Accelerating preliminary eco-innovation design for products that integrates case-based reasoning and TRIZ method. J. Clean. Prod. **19**(9–10), 998–1006 (2011). https://doi.org/10.1016/j.jclepro.2011.01.014

Yang, C.J., Chen, J.L.: Forecasting the design of eco-products by integrating TRIZ evolution patterns with CBR and Simple LCA methods. Expert Syst. Appl. **39**(3), 2884–2892 (2012). https://doi.org/10.1016/j.eswa.2011.08.150

Critical Success Factors on ERP Implementations: A Bibliometric Analysis

Pau Vicedo[1(✉)], Hermenegildo Gil[1], Raúl Oltra-Badenes[1], and José M. Merigó[2]

[1] Department of Business Organisation, Universitat Politècnica de València, Camí de Vera s/n, 46022 València, Spain
pavipa@upv.es
[2] Department of Management Control and Information Systems, University of Chile, Diagonal Paraguay, 205 – 2057, 8330015 Santiago, Chile

Abstract. This work presents a bibliometric analysis of the influential authors, institutions, papers and countries on the field of Enterprise Resource Planning implementations and their Critical Success Factors based on the Web of Science database with 301 articles belonging to 86 Universities and Institutions from 48 different countries. The research has been conducted from 1999 to end of 2017.

Keywords: CSF · ERP · Critical Success Factors · Web of Science · Enterprise Resource · Planning

1 Introduction

In a global and digital world all Small and Medium-sized Enterprises (SMEs) face new situation where they must respond quicker to market changes. In order to do that and persist in time new systems have to be implemented and this is the case of the Enterprise Resource Planning Systems (ERP) which are a critical tool to react to the above situation. This kind of systems are able to provide real-time information about the company in all aspects such as financial, production, planification among many others (Bradford et al. 2003) making decision making easier and quicker to adapt to market changes. However, not just implementing an in-house ERP is enough for an SME to whether survive, grow or just to reach their goals.

Over the past two decades, many investigations have identified a number of Critical Success Factors (CSF) which, if followed, significantly increase the success rate of ERP implementation (Hedman 2010). These CSF, are defined as a necessary fact or element to achieve one mission. Thus, it is crucial to ensure and identify that the implementation covers them as a critical part for a successful ERP implementation.

Following Broadus (1987), bibliometric is defined as the discipline that studies the bibliographic material quantitatively. Starting from their inception, ERP systems and their CSF for successful implementations have produced a substantial number of researches that led to several publications and studies (Hsieh and Chang 2009). The goal of this work is to identify the most influential articles, authors, institutions, countries and journals in the existing scientific bibliography through a bibliometric

J. C. Ferrer-Comalat et al. (Eds.): MS-18 2018, AISC 894, pp. 169–181, 2020.
https://doi.org/10.1007/978-3-030-15413-4_13

study, from a general perspective, taking into account different indexes that help to provide a better understanding of the state of the art and also provide trends in the field of study.

Although Web of Science (WoS) is not the only Scientific database, it is the reference database in scientific research contributions which is the core of this work, however, WoS is a live database which is continuously growing and adding more publications constantly, for that reason it has been taken a static picture of the existing publications by January 2018 and only considering the complete natural years. This study provides an overview of the evolution of published studies during time, the most influential 20 papers, the most influential universities and authors, the most prominent countries and, finally, the journals publishing on this field, considering for all of them both citations over years and publications. Previously to the results there is a section exposing the method used and following there is a conclusion summary.

2 Methodology

Studying and reviewing the literature on a specific topic gives an overview of tendencies of research in it and its impact (Anguinis et al. 2012; Pilkington and Meredith 2008). The current study considered all the works over the years published on Web of Science (WoS), however only articles and reviews since 1999 have been found and only published studies during complete natural years have been considered. Although, there are other Academic databases, such as Google Scholar, Scopus, etc., WoS is mainly considered the best academic database for studying research contributions (Shiau 2015) it contains more than 15,000 journals, over 90,000,000 records and it covers up to 273 disciplines.

Based on other bibliometric studies (Fagerberg 2012; Dereli et al. 2011; García Merino et al. 2016), a wide variety of methods are used but the most common indicators are the number of citations and the amount of publications. In this study, those two indicators and their relation have been considered together with the H-index, a newly index introduced by Hisrch (2005), that integrates publications and citations into one single index, all of them define the impact of each article (Stonebraker 2012). Moreover, as Merigó et al. (2015) stated, the number of articles above a citation threshold permits to identify influential articles, reason why different thresholds are considered on the results. Additionally, it is important to identify the institutions quality (Podsakoff et al. 2008) because that indicate the importance and impact of an article as well. Finally, a study about which Journals (Goh et al. 1997) are more influential about the research topic is conducted to complete the whole picture about the state of the art.

For conducting the work, since WoS database is enormous, some limitations have been applied in order to narrow the results to the specific field of study. A selection of keywords related to the topic was considered: ERP, CSF, Enterprise Resource(s) Planning, Critical Success Factor(s) and Enterprise System(s). That generated 549 references but not all of them were related to the main topic, consequently not all categories have been taken into account, only those within the field of study: computer science information systems, management, information science library science, operations research management science, computer science interdisciplinary applications, business, engineering industrial, computer science theory methods, engineering manufacturing, computer science artificial intelligence, engineering electrical electronic, economics, computer science software engineering, engineering multidisciplinary, business finance, computer science hardware architecture, telecommunications, social sciences interdisciplinary, computer science cybernetics, automation control systems, planning development, engineering mechanical and ergonomics reducing the amount of publications to 491 and then only the article (287 papers) and reviews (15) are taken, making an amount of 301 works.

The study uses the material available on WoS in January 2018.

3 Results

In this section we will present the most significant results of the bibliometric analysis by different approaches. Following the mentioned keywords and related categories of study, 491 references were found including 287 articles, 17 book chapters, 199 proceedings papers, 15 reviews, 2 editorial materials and 1 book. Those works have a total amount of citations of 8999 with a ratio (cites/studies) of 18.33, the h-index is 44.

For reducing the results to some more accurate quality materials only the articles and the reviews have been considered making an overall of 301 references, since those options are the ones which can be admitted as pure scientific contributions. Taking into consideration that result we obtain 8817 citations with a ratio (cites/studies) of 29.29 and a h-index of 44.

3.1 Evolution of Published Studies

The first article of which there is record was published in 1999, being the only publication of that year. Figure 1 shows graphically that the amount of studies increased slowly over years, reaching 26 works in 2008 with the exception of 2006 where only 5 studies were published. After 4 years of decline in the number of publications, 2015 and 2016 where the most prolific years with 35 publications followed by a minor descent in 2017. The gradual increase of publications it is due to two main factors, the first one is the increase of researchers worldwide, the second and main reason is the digitalization of society. Increasingly, we see that companies need more up-to-date information to make decisions faster and to be able to adapt to the market changes, there is where the ERP systems play a critical role. The scientific community is aware of the world digitalization and invests more resources into this field.

Analyzing the citations in Table 1. General citation per year, it can be seen that the most cited articles are not the ones lately published. The year 2007 was the year with the most citations with 1464, other years with more than 1000 citations were 2003, 2005 and 2008. This is not related to the number of articles published which increase almost every year but due to the fact that those article have been published longer.

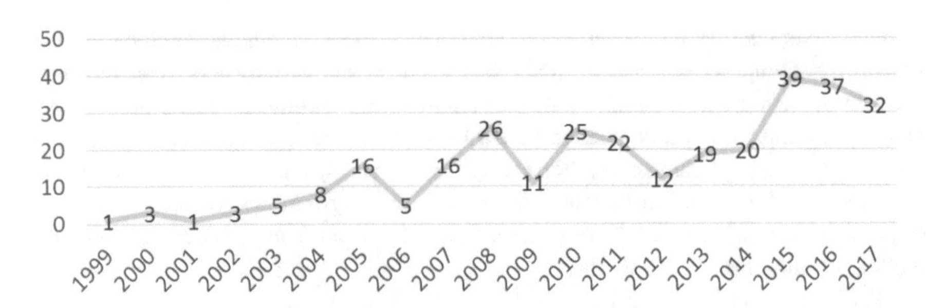

Fig. 1. Published articles

Table 1. Cited studios per year

Year	≥ 50	≥ 10	≥ 1	TS	TC
1999	1	0	0	1	351
2000	2	0	1	3	243
2001	0	1	0	1	35
2002	3	0	0	3	776
2003	3	1	1	5	1009
2004	5	2	0	8	725
2005	8	4	3	16	1120
2006	2	3	0	5	243
2007	5	9	1	16	1470
2008	6	17	3	26	1126
2009	2	4	4	11	288
2010	2	13	8	25	467
2011	0	7	11	22	188
2012	0	7	4	12	165
2013	2	8	7	19	263
2014	0	4	15	20	151
2015	0	1	29	39	126
2016	0	0	19	37	53
2017	0	0	9	32	18
Total	41	81	115	301	8817
PCT	13,62	26,91	38,21	100	

Abbreviations: ≥ 50 = Number of documents with equal or more than 50 citations; ≥ 10 = Number of documents with equal or more than 10 citations; ≥ 1 = Number of documents with equal or more than 1 citations; TS = Total studies; TC = Total citations; PCT = Percentage.

3.2 Most Influential Papers

In order to measure the impact of an article it is important to identify not only how many times it has been cited but also when it was published. The combination of those both indexes give a clear indication of the impact of each work. On Table 2 a list of most influential publications can be seen. On top of it, Liang, Huigang; Saraf, Nilesh; Hu, Qing et al. (2007), has a ratio of 74,27 citations per year and it is also the most cited article with 817 citations, followed by Umble, EJ; Haft, RR; Umble, MM (2003) with 513 citations and average of 34.20 citations per year.

There are 20 papers over 10 citations per year and all of them but one belong to the 21st century which indicates that there are more and more researches being conducted on this field. It is important to highlight that these publications are heavily cited.

Table 2. Most influential articles

R	C/Y	Title Author/s	TC	Year
1	74,27	Assimilation of enterprise systems: The effect of institutional pressures and the mediating role of top management Liang, Huigang; Saraf, Nilesh; Hu, Qing; et al.	817	2007
2	34,27	Enterprise resource planning: Implementation procedures and critical success factors Umble, EJ; Haft, RR; Umble, MM	514	2003
3	27,13	The critical success factors for ERP implementation: an organizational fit perspective Hong, KK; Kim, YG	434	2002
4	22,00	Enterprise resource planning: A taxonomy of critical factors Al-Mashari, M; Al-Mudimigh, A; Zairi, M	330	2003
5	18,47	A critical success factors model for ERP implementation Holland, CP; Light, B	351	1999
6	18,46	What happens after ERP implementation: Understanding the impact of interdependence and differentiation on plant-level outcomes Gattiker, TF; Goodhue, DL	240	2005
7	17,70	Examining the critical success factors in the adoption of enterprise resource planning Ngai, E.W.T.; Law, C.C.H.; Wat, F.K.T.	177	2008
8	17,06	Vicious and virtuous cycles in ERP implementation: a case study of interrelations between critical success factors Akkermans, H; van Helden, K	273	2002
9	15,93	A taxonomy of players and activities across the ERP project life cycle Somers, TM; Nelson, KG	223	2004
10	15,31	A framework of ERP systems implementation success in China: An empirical study Zhang, Z; Lee, MKO; Huang, P; et al.	199	2005

(*continued*)

Table 2. (*continued*)

R	C/Y	Title Author/s	TC	Year
11	13,27	Risk management in ERP project introduction: Review of the literature Aloini, Davide; Dulmin, Riccardo; Mininno, Valeria	146	2007
12	12,07	Enterprise information systems project implementation: A case study of ERP in Rolls-Royce Yusuf, Y; Gunasekaran, A; Abthorpe, MS	169	2004
13	11,38	Identifying critical issues in enterprise resource planning (ERP) implementation Ehie, IC; Madsen, M	148	2005
14	11,20	TAM-based success modeling in ERP Bueno, Salvador; Salmeron, Jose L	56	2013
15	11,00	A grey-based DEMATEL model for evaluating business process management critical success factors Bai, Chunguang; Sarkis, Joseph	55	2013
16	10,82	The impact of ERP implementation on business process outcomes: A factor-based study Karimi, Jahangir; Somers, Toni M.; Bhattacherjee, Anol	119	2007
17	10,50	Organizational culture and leadership in ERP implementation Ke, Weiling; Wei, Kwok Kee	105	2008
18	10,25	A multi-project model of key factors affecting organizational benefits from enterprise systems Seddon, Peter B.; Calvert, Cheryl; Yang, Song	82	2010
19	10,22	A model of ERP project implementation Parr, A; Shanks, G	184	2000
20	10,20	Implementation critical success factors (CSFs) for ERP: Do they contribute to implementation success and post-implementation performance? Ram, Jiwat; Corkindale, David; Wu, Ming- Lu	51	2013

Abbreviations available in Tables 1 and 2 except for R = Rank and C/Y = Citations per Year

3.3 Most Influential Institutions

Table 3 shows the 20 most productive institutions worldwide, in this case the number of publications, times cited, h-index and the ratio (times cited/total studies) are considered together with the most relevant number of papers above the given thresholds 50 and 20. This way the lector can have an idea of the real influence of each university. The top influential Universities are Florida Atlantic University from United States, University of Manchester from United Kingdom, Eindhoven University of Technology from the Netherlands, University of Melbourne from Australia and Boise State University from United States, all of them above 130 citations per publication and above 265 overall citations. Regarding the production, the City University of Hong Kong and the Monsah University from Australia are the most productive with 5 publications each one with an h-index of 3, which means they have 3 articles with, at least, 3 citations each. Following are the Wayne State University and University of Nebraska Lincoln from United States and University of Sheffield from United Kingdom.

The universities from the United States are the most listed with 10 appearances followed by United Kingdom (which includes the works published under England, Scotland, Wales and Northern Ireland) with 4, Australia with 3, China 2 appearances and the Netherland with just one. Even though the most influential institutions are from English speaking countries such as US, UK, Canada and Australia it can be stated that European and Asian Universities are growing. This is due to the rising of the Asian universities which are committed to the technology research and how to apply it to the business environment. However, most of the universities are from English speaking countries. In Fig. 2 we can see the relationship between some institutions.

Table 3. Most influential institutions

R	Institution	Country	TS	TC	H	TC/TS	≥50	≥20
1	Florida Atlantic U	US	3	1006	3	335,33	3	0
2	U of Manchester	UK	2	382	2	191,00	1	1
3	Eindhoven U of Technology	NL	2	287	2	143,50	1	0
4	U of Melbourne	AU	2	266	2	133,00	2	0
5	Boise State U	US	2	265	2	132,50	1	1
6	Accenture	US	2	213	2	106,50	2	0
7	Wayne State U	US	4	415	4	103,75	3	0
8	U of South Florida	US	2	187	2	93,50	2	0
9	U of Hull	UK	2	171	2	85,50	1	0
10	U of Nebraska Lincoln	US	4	328	4	82,00	2	2
11	U of Leeds	UK	2	154	2	77,00	2	0
12	U of Colorado Denver	US	3	189	2	63,00	2	0
13	City U of Hong Kong	CN	5	310	3	62,00	2	0
14	U of Colorado Health Science Center	US	2	121	2	60,50	1	0
15	Kansas State U	US	4	228	3	57,00	2	0
16	Monash U	AU	5	273	3	54,60	2	0
17	Hong Kong Polytechnic U	CN	4	216	3	54,00	1	1
18	U of Sheffield	UK	4	175	4	43,75	1	2
19	Old Dominion U	US	2	80	2	40,00	2	0
20	U of Southern Queensland	AU	3	107	3	35,67	1	1

Abbreviations available in Table 1 except for R = Rank and H = h-index

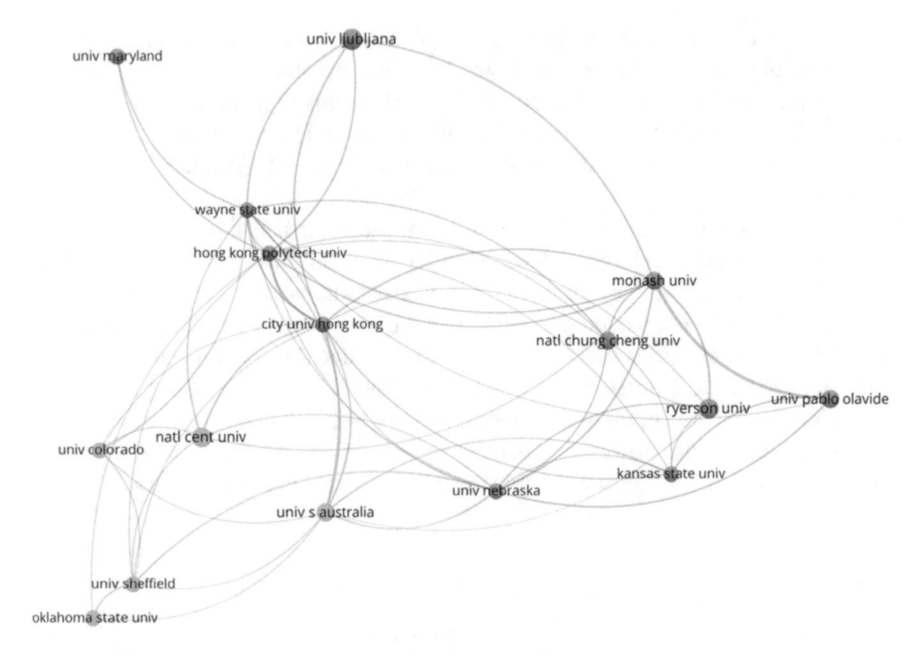

Fig. 2. Relationship between Institutions

3.4 Most Influential Authors

As well as the institutions, there are authors publishing about ERP and their CSF for successful implementations from all over the world. Table 4 presents the 20 most influential authors ordered by citations per year (TC/TS). As the other sections, the h-index, the times cites and total studies appear to get a better picture of their production and influence. Table 4 also includes the University/Institution where their last paper was submitted to WoS and their country.

As shown in the Table 4, Americans are the most influential authors with 8 out of top 10, Liang and Xue share the first position as they co-authored the most cited paper with 930 citations with an index of 465 citations per year. The first not American author is the British Abthorpe, MS with a ratio of 169 citations per article with just one article, followed by the Chinese Law with 107 citations per work with 214 citations overall. English-speaking authors from US and UK occupy 18 of the top 20 places in the list emphasizing the importance of Anglo-Saxon countries in this field of research. Also noteworthy are the Spanish authors placing 1 in the top 20 being one of the most productive with a published number of papers and the University with more influential authors are Old Dominion University from United States with 3, Florida Atlantic University and Kansas State University from United States and University of Leads from United Kingdom with two authors each one.

Table 4. Most influential authors

R	Name	Institution	Country	TS	TC	TC/TS	≥ 50	≥ 20	H
1	Liang, HG	Florida Atlantic U.	United States	2	930	465,00	2	0	2
2	Xue, YJ	Florida Atlantic U.	United States	2	930	465,00	2	0	2
3	Hu, Q	Iowa State U.	United States	3	922	307,33	2	1	3
4	Abthorpe, MS	Nottingham Trent U.	United Kingdom	1	169	169,00	1	0	1
5	Somers, TM	Wayne State U.	United States	3	410	136,67	3	0	3
6	Gattiker, TF	Boise State U.	United States	2	265	132,50	1	1	2
7	Law, CCH	Hong Kong Polytechnic U.	China	2	214	107,00	1	1	2
8	Gunasekaran, A	U. of Massachusetts Dartmouth	United States	2	194	97,00	1	1	2
9	Bhattacherjee, A	U. of South Florida	United States	2	187	93,50	2	0	2
10	Karimi, J	U. of Colorado Denver	United States	2	187	93,50	2	0	2
11	Burgess, TF	U. of Leads	United Kingdom	2	154	77,00	2	0	2
12	King, SF	U. of Leads	United kingdom	2	154	77,00	2	0	2
13	Nah, FFH	Missouri U. of Science & Technolofy	United States	4	285	71,25	2	1	3
14	Koh, SCL	U. of Sheffield	United Kingdom	2	138	69,00	1	1	2
15	Li, L	Old Dominion U.	United States	2	80	40,00	0	2	2
16	Markowski, C	Old Dominion U.	United States	2	80	40,00	0	2	2
17	Xu, L	Old Dominion U.	United States	2	80	40,00	0	2	2
18	Sheu, C	Kansas State U.	United States	2	79	39,50	1	0	2
19	Yen, HR	Kansas State U.	United States	2	79	39,50	1	0	2
20	Bueno, S	U. Pablo de Olavide	Spain	3	99	33,00	1	1	2

3.5 Most Influential Countries

Being the field of study a generic field for research that affects any SME all over the world and not specific to any geographical area, there are published articles from 48 different countries in the WoS database. Table 5 presents the top 20 influential countries. Same indexes than the ones used on the previous tables are used here as well. Note that the countries listed refer to the country where the author was working at the time of the publication which is not necessarily their country of origin.

Surprisingly the most influential country is South Korea with a ratio of 73.88 citations per work due to the high number of citations of their only 8 publications. It is followed by Canada with a ratio of 61.44 but with many more works published, 1106. In third place is Netherlands with a ratio of 60.6 and only 5 publications. As expected for the quality of their works, and already stated by the reviewed literature (Merigó et al. 2016), the importance of their institutions and its size the US is the number one in works (77) and times cited (4083) with an h-index of 29 but far from being the most influential country in citations per study.

Nonetheless, being the most productive country is not guarantee of being the most influential. Even so, the Anglo-Saxon countries occupy 3 of the first 6 positions, which reaffirms what has been studied in the previous sections that they are the ones that,

Table 5. Most influential countries

R	Country	TS	TC	H	TC/TS	≥ 50	≥ 20
1	South Korea	8	591	6	73,88	2	1
2	Canada	18	1106	9	61,44	2	6
3	Netherlands	5	303	4	60,60	1	0
4	USA	77	4083	29	53,03	20	19
5	Saudi Arabia	7	367	4	52,43	1	0
6	United Kingdom	34	1613	16	47,44	7	8
7	China	20	736	10	36,80	5	3
8	Bangladesh	1	27	1	27,00	0	1
9	Ireland	1	25	1	25,00	0	1
10	Australia	23	561	10	24,39	4	3
11	Italy	8	186	5	23,25	1	0
12	Switzerland	4	86	2	21,50	1	0
13	Singapore	5	105	4	21,00	1	0
14	Taiwan	31	568	13	18,32	3	6
15	Malaysia	5	91	3	18,20	1	1
16	Norway	3	48	2	16,00	0	1
17	Spain	15	210	6	14,00	1	3
18	Finland	3	38	1	12,67	0	1
19	France	7	85	4	12,14	0	2
20	New Zealand	5	59	2	11,80	0	1

in general, influence the rest of the investigations the most. If we look at the origin by continents, we can see that Europe is the continent with more countries in the list of the most influential with 9, followed by Asia with 7 and, finally, America and Oceania with two countries each. It is worth highlighting that 68,4% of published studies come from English speaking countries event though they only occupy three 1 position on the top 3.

3.6 Publishing Journals

In this section an overview of the journals (Petersen et al. 2011; Maloni et al. 2012) publishing on this field is given. In Table 6 can be seen which the Journals with higher ratio of publications per study on this field are. At the top of the list MIS Quarterly with a ratio of 381 citations per study but with only 3 studies published, followed by IEEE Software with 351 citations for the only article they have ever published and European Journal of Operational Research in third position with a ratio of 288.67 and 3 publications.

In the other hand, the list varies in case of looking at the number of published studies. In this case International Journal of Production Economics is on top with 14 studies followed by Computer in Industry and Journal of Computer Information Systems with 12 publications each Journal.

Looking at the overall citations per Journal, MIS Quarterly (1143) is again in the first position but, in this case, it is followed by Information Management (965), European Journal of Operational Research (866) and International Journal of Production Economics (864).

Table 6. Most influential journals

R	Name	TS	TC	H	TC/TS	≥ 50	≥ 20
1	MIS Quaterly	3	1143	3	381,00	3	0
2	IEEE Software	1	351	1	351,00	0	0
3	European Journal of Operational Research	3	866	3	288,67	2	1
4	Information Management	9	965	7	107,22	4	2
5	Journal of Information Technology	2	184	1	92,00	1	0
6	Industrial Marketing Management	1	84	1	84,00	1	0
7	International Journal of Human Computer Interaction	2	141	1	70,50	1	0
8	International Journal of Production Economics	14	864	12	61,71	7	2
9	Interacting with Computers	1	56	1	56,00	1	0
10	European Journal of Information Systems	9	473	8	52,56	2	3
11	Journal of Management Information Systems	4	201	3	50,25	2	0
12	IEEE Transactions on Systems Man and Cybernetics Part C Applications and Reviews	1	49	0	49,00	0	1
13	Information Systems	1	43	1	43,00	0	1
14	Journal of Systems and Software	2	79	2	39,50	0	2
15	Supply Chain Management an International Journal	2	77	2	38,50	1	0
16	Advances in Engineering Software	1	36	1	36,00	0	1
17	Decision Support Systems	4	143	3	35,75	1	0
18	Computers in Industry	12	422	7	35,17	2	1
19	Industrial Management Data Systems	8	277	6	34,63	3	1
20	Journal of Computer Information Systems	12	407	7	33,92	3	3

4 Conclusions

This study gives a general overview of the works published about the ERP implementation and its CSF on the WoS. The results show how the number of publications has increased over the years from 1 in 1999 to more than 30 since 2015 due to the continuous business digitalization and the necessity of data to make faster decisions. US lead the publications in terms of productivity (overall published articles), influential authors and influential institutions because they host the main Universities. However, due to their high number of publications they are not at the top of the most influential country if the ratio published studies/times cited is taken into consideration. Asian and European countries follow closely, it is worth to highlight the importance of South Korea and Netherland in this field with a strong position and important ratio of publications/citations. In the successive positions we see other English-speaking countries such as Australia, United Kingdom and Canada. Developing countries do not appear on the leading positions but they start to publish also on this field.

Regarding the Journals, it is important to highlight that the most cited ones are not the most productive. The most cited article was published in MIS Quarterly which is one of the most important Journals in the field. Moreover, the signing authors of that article, Liang, HG and Xue, YJ, are also considered the most influential and their university, Florida Atlantic University, is also the first one on institutions. Therefore, their article is considered the most important article with most impact on all other researches.

References

Bradford, M., Florin, J.: Examining the role of innovation diffusion factors on the implementation success of enterprise resource planning systems. Int. J. Acc. Inf. Syst. **4**(3), 205–225 (2003)

Hedman, J.: ERP systems: Critical factors in theory and practice. C Center for Applied ICT (CAICT), CBS, Frederiksberg (2010)

Broadus, R.: Toward a definition of "bibliometrics". Scientometrics **12**(5–6), 373–379 (1987)

Hirsch, J.E.: An index to quantify an individual's scientific research output. Proc. Nat. Acad. Sci. U.S.A. **102**(46), 16569 (2005)

Merigó, J.M., Gil-Lafuente, A.M., Yager, R.R.: An overview of fuzzy research with bibliometric indicators. Appl. Soft Comput. **27**, 420–433 (2015)

Anguinis, H., Suárez-González, I., Lannelongue, G., Joo, H.: Scholarly impact revisited. Acad. Manag. Perspect. **26**(2), 105–132 (2012)

Dereli, T., Durmuşoğlu, A., Delibaş, D., Avlanmaz, N.: An analysis of the papers published in total quality management & business excellence from 1995 through 2008. Total Qual. Manag. Bus. Excell. **22**(3), 373–386 (2011)

Petersen, C.G., Aase, G.R., Heiser, D.R.: Journal ranking analyses of operations management research. Int. J. Oper. Prod. Manag. **31**(4), 405–422 (2011)

Maloni, M., Carter, C.R., Kaufmann, L.: Author affiliation in supply chain management and logistics journals: 2008–2010. Int. J. Phys. Distrib. Logist. Manag. **42**(1), 83–100 (2012)

Hsieh, P.N., Chang, P.L.: An assessment of world-wide research productivity in production and operations management. Int. J. Prod. Econ. **120**, 540–551 (2009)

García Merino, M.T., Pereira do Carmo, M.L., Santos Álvarez, M.V.: 25 years of technovation: characterisation and evolution of the journal. Technovation **26**, 1303–1316 (2016)

Podsakoff, P.M., MacKenzie, S.B., Podsakoff, N.P., Bachrach, D.G.: Scholarly influence in the field of management: a bibliometric analysis of the determinants of University and author impact in the management literature in the past quarter century. J. Manag. **34**, 641–720 (2008)

Goh, C.H., Holsapple, C.W., Johnson, L.E., Tanner, J.R.: Evaluating and classifying POM journals. J. Oper. Manag. **15**, 123–138 (1997)
Pilkington, A., Meredith, J.: The evolution of the intellectual structure of operations management —1980–2006: a citation/co-citation analysis. J. Oper. Manag. **27**, 185–202 (2008)
Stonebraker, J.S., Gil, E., Kirkwood, C.W., Handfield, Robert B.: Impact factor as a metric to assess journals where OM research is published. J. Oper. Manag. **30**, 24–43 (2012)
Fagerberg, J., Fosaas, M., Sapprasert, Koson: Innovation: exploring the knowledge base. Res. Policy **41**, 1132–1153 (2012)
Shiau, W.L., Dwivedi, Y.K., Tsai, Chia-Han: Supply chain management: exploring the intellectual structure. Sciencentometrics **105**, 215–230 (2015)
Merigó, J.M., Cancino, C.A., Coronado, F., Urbano, D.: Academic research in innovation: a country analysis. Sciencentometrics **108**, 559–593 (2016)

A Bibliometric Analysis of Leading Countries in Supply Chain Management Research

Keivan Amirbagheri[1(✉)], José M. Merigó[2,3], and Jian-Bo Yang[4]

[1] Department of Business Administration, University of Barcelona, Av. Diagonal 690, 08034 Barcelona, Spain
keivan.amirbagheri@gmail.com

[2] Department of Management Control and Information Systems, School of Economics and Business, University of Chile, Av. Diagonal Paraguay 257, 8330015 Santiago, Chile

[3] School of Information, Systems and Modelling, Faculty of Engineering and Information Technology, University of Technology Sydney, 81 Broadway, Ultimo, NSW 2007, Australia

[4] Manchester Business School, University of Manchester, Booth Street West, Manchester M15 6PB, UK

Abstract. Supply chain management as a newly comer discipline has attracted many attentions of the scholars to do an investigation based on its prominent level of importance for the economy and its influence on the management of the organizations. So, the key point is to understand the trends among the countries throughout the time to have a powerful insight about this issue. To this end, this work does a comprehensive analysis from 1990 to 2017. The purpose of this study is to analyze the leading countries and understand thoroughly their trends during the time. The work has dedicated to three sections. In the first one the countries have studied globally to give a comprehensive overview to academics. Next, the performance of the countries is studied in three periods to understand better the changes of each during the time. Finally, some individual journals and groups of journals are also investigated. The results show that the USA is the leader of the countries while China has experienced an enormous growth and it is predictable that with this trend can reach to the top of the list.

Keywords: Supply chain management · Production and operations management · Bibliometrics · Web of Science · VOS viewer · Country

1 Introduction

Supply Chain Management (SCM) as a new concept has been presented for the first time by Oliver and Webber (1982) and they consider it as a strategy to integrate the activities of a supply chain. Although the Supply Chain Management (SCM) concept was born at the beginning of the 1980s, research in the field was almost non-existent until the mid-1990s (Alfalla-Luque and Medina-López 2009). Since then this concept has received a lot of attention from scholars and day by day its scope is developing and nowadays SCM has developed into a multivariate discipline (Cousins et al. 2006).

J. C. Ferrer-Comalat et al. (Eds.): MS-18 2018, AISC 894, pp. 182–192, 2020.
https://doi.org/10.1007/978-3-030-15413-4_14

Bibliometric analysis as an efficient tool could be applied to classify the numerous publications in SCM. There are some works in this discipline that have analyzed SCM with the context of bibliometric analysis. Charvet et al. (2008) use a bibliometric approach to study the intellectual structure of supply chain management. In another work, Alfalla-Luque and Medina-López (2009) examine SCM and its influence on the needs of companies by analyzing the bibliometric studies of the main journals in the discipline. Although the existence of these articles and each of them analyze the concept of SCM generally but the lack of a study with the concentration on the leading countries and their productivity and influence also obtaining a comprehensive overview in SCM seems necessary. Additionally, bibliometric analysis as a powerful tool in many articles. As an example, Merigó and Yang (2017) present a bibliometric overview of research published in operations research and management science in recent decades. Cancino et al. (2017) develop a bibliometric analysis of the publications of the Computers and Industrial Engineering between 1976 and 2015 to identify the leading trends of the journal in terms of impact, topics, universities and countries. In the very same work, Martínez-López et al. (2018) provide a bibliometric overview of the leading trends of the European Journal of Marketing between 1967 and 2017. In a similar work Merigó et al. (2018) use a bibliometric method to identify the most relevant authors, institutions, countries, and analyze their evolution through time in Information Sciences. In another anniversary paper using bibliometric overview, Leangle et al. (2017) present a general overview of the journal over its lifetime by using bibliometric indicators. Finally, Merino et al. (2006) analyze the characterization and evolution of the Technovation journal in a period of 25 years.

From the other point of view there are many studies around production and operations management, as an important discipline, that apply bibliometric methods or some other similar methods. Smith et al. (2008) examine the institutional factors that affect the productivity of individuals in the field of operations by using a partial least squares analysis. In another work the ranking of journals for OM research using meta-analysis is done (Petersen et al. 2011). Hsieh and Chang (2009), through papers published in 20 core POM journals assess the global performance of POM research. In a similar work Goh et al. (1997) evaluate and classify POM journals in their work also Barman et al. (2001) analyze the perceived relevance and quality of POM journals for 10 years. The other relevant work is the work of Theoharakis et al. (2007) that they provide peer review evaluations for POM research outlets. Stonebraker et al. (2012) investigate impact factor as a metric for ranking the quality of journal outlets for operations management research. Holsapple and Lee-Post (2010) do a behavior-based analysis of knowledge dissemination channels in operations management, whilst Lindermand and Chandrasekaran (2010) study the exchange of ideas within Operations Management journals and between other management disciplines (Management, Marketing, and Finance) during the (1998–2007). Fry and Donohue (2013) do a data envelopment analysis assessment of journal quality and rankings for operational management research. Pilkington and Meredith (2009) study the evolution of the intellectual structure of operations management between 1980 and 2006 through a citation/co-citation analysis.

The main objective of this work is to analyze the most important and the most influential countries in SCM through analyzing them with bibliometric measures. To do so, this work first presents a global overview through a comprehensive table and tries to report what is going on in SCM research area. Second, by dividing the period of almost 30 years (1990–2017) to three periods the authors seek the trend of each period for each country and to follow up the probable changes during these three periods. Through this work the authors try to give a thorough understanding about the pioneer countries in SCM from diverse points of view. At the first glance to the general table that reports the global trend of the publications since the very first year, it is noticeable that the USA, China and UK are on the top of the leading countries not only from the total number of publications point of view, but also from the total citation of each country.

2 Bibliometric Methods

In the world of academic research, there are a plenty of databases that can lead an investigator toward a high-quality search. Among them Web of Science (WoS) as a prestigious database that classifies the published articles in the journals with the highest quality is the option of the authors and with the aim of access the world's leading scholarly literature in the sciences, social sciences, arts, and humanities and examine proceedings of international conferences, symposia, seminars, colloquia, workshops, and conventions, Web of Science Core Collection is the selected database for beginning the process of search. Selecting the appropriate keywords is essential part of the search that can lead us to the best possible results. So, to cover all the publications in this discipline, the authors use "supply chain" or "supply chain management" or "SCM" from one side and from the other side some important journals in SCM area that are: "Supply Chain Management: an International Journal or Journal of Supply Chain Management or International Journal of Information Systems and Supply Chain Management or Journal of Humanitarian Logistics and Supply Chain Management or Journal of Transport and Supply Chain Management or Operations and Supply Chain Management an International Journal" between 1990 to 2017. The reason behind this decision is to not neglect from the papers that could be useful to improve the quality of this work.

Based on this search the work obtains 35497 articles. But to have more accurate results filtering some categories seems necessary. These categories are: Operations Research Management Science, Management, Engineering industrial, Manufacturing and Environmental, Business and Business Finance, Economics, Agriculture Economics Policy and Computer Science. The next step to achieve to research studies, is to refine articles, reviews, letters and notes according to the WoS categories. So, the final number of the articles is 20616 papers. The entire process of search was done on March 2018.

Bibliometrics is the research field of library and information sciences that studies the bibliographic material by using quantitative methods (Broadus 1987; Pritchard 1969). This concept sheds a light on the research topic and makes it easier to report the obtained results. Generally, there are various indicators to evaluate the information such as number of publications, citations (Shiau et al. 2015) and *h-index* (Hirsch 2005). The *h-index* is a combination of number of publications and citations such that for example if the *h-index* of a group of publications is 20 means that 20 papers of the group have received 20 or more citations. In this study for each item all these indicators have been calculated but the ranking is based on total publications of each of them.

Although the search results show that before 1990 there are a few numbers of published papers but since 1990 the number of the publications in SCM has experienced an increasing trend so the bibliometric results are organized from this year so on and around country analysis. First, the countries have been studied globally to understand their individual role and their influence in the field of SCM research in a global context. Second classification analyzes the countries periodically. Hence, there are three periods: "1990–1999", "2000–2009" and "2010–2017". The purpose of this analysis is exactly like the global one but to look up the leading countries during each period. Finally, the study analyzes the important journals in this area with an individual approach (Merigó et al. 2015) to study the contribution of each country in the leading journals of the field and some other important groups of journals. The logic behind this classification is the role of these journals.

VOS viewer is a powerful software that is used to report visually the countries of the study through bibliographic coupling (Van Eck and Waltman 2010).

3 Results

3.1 Leading Countries in Supply Chain Management

One of the most interesting and most important outputs that can be analyzed, is the analysis of the countries and their role and their efficiency in the world of SCM. To do so, the authors have analyzed the productivity and influence of the top 50 leading countries in this area between 1990 and 2017. The ranking of the countries is based on total publications of each of them but there is some extra information to present the analysis as comprehensive as possible like the number of citations, the number of papers respectively with more than 250, 100 and 50 citations. Table 1 reports briefly all these items for the top 50 countries in SCM. Among them the USA, China and UK are the most productive countries of the world in SCM.

It is noteworthy to mention the presence of some other Asian countries like India, Taiwan and Iran among the first top 15 countries of this list. Although they still don't play a significant role in the world of SCM, but their presence is prominent. Besides this, the countries that have occupied the top ranking were expectable to be in these positions.

Besides, VOS viewer software and the graphics which produces could give a profound insight about the outputs and makes us able to analyze them visually. So, through this software we extract the bibliographic coupling of countries. Figure 1 demonstrates all the countries in this field according to the citation number with a decreasing trend. The characteristics of the figure are the threshold of 96 countries and the 100 most representative connections.

Because the used data to create these graphics is the primitive data of WoS so England, Wales, Scotland and North Ireland from one side and China and Hong Kong from the other side, has been displayed distinctly.

Table 1. Most influential countries in SCM research

R	Country	TPSCM	TCSCM	≥ 1000	≥ 500	≥ 250	≥ 100	≥ 50
1	USA	6,789	217,514	3	23	108	474	1,126
2	China	3,305	57,043	0	2	13	85	263
3	UK	2,368	58,233	0	5	11	103	310
4	Taiwan	1,200	20,803	0	1	1	31	93
5	Canada	1,053	27,221	0	0	13	58	125
6	Germany	1,042	22,063	1	3	7	40	99
7	India	1,010	15,048	0	1	2	22	65
8	Italy	741	14,607	0	0	1	20	72
9	Australia	734	12,413	0	1	2	16	58
10	France	731	14,022	0	1	6	25	61
11	Iran	726	9,628	0	0	1	10	39
12	Netherlands	645	14,750	0	1	6	22	65
13	Spain	634	9,801	0	0	0	15	38
14	South Korea	583	9,807	0	0	0	19	50
15	Sweden	495	8,314	0	0	0	12	36
16	Turkey	427	8,885	0	0	2	13	48
17	Denmark	339	8,396	0	1	2	18	42
18	Singapore	332	7,611	0	1	2	13	36
19	Brazil	318	3,301	0	0	0	5	11
20	Finland	316	4,777	0	0	0	5	22
21	Japan	299	5,285	0	1	2	7	18
22	Switzerland	276	6,026	0	1	1	10	28
23	Malaysia	229	3,321	0	0	1	6	17
24	Belgium	215	5,410	0	0	3	11	29
25	Greece	211	4,634	0	0	0	7	31

TCSCM, TPSCM, number of citations and studies in supply chain management; >1000, >500, >250, >100, >50, number of articles with more than 1000, 500, 250, 100 and 50 cites.

As it is clear from this figure, the USA is the most productive country and has the biggest bibliographic coupling structure. Besides, it is noteworthy that England, India and China also achieve an appropriate position in this context.

3.2 Periodical Analysis of Leading Countries

One of the potentially interesting issues that can be analyzed is doing a periodic analysis of the countries in SCM. The aim of this analysis is to study the trend of each country during different periods and observe their increasing, decreasing or monotone movements. These changes during the time may have several reasons, maybe countries' economic situation or perhaps the tendency of countries on this discipline based

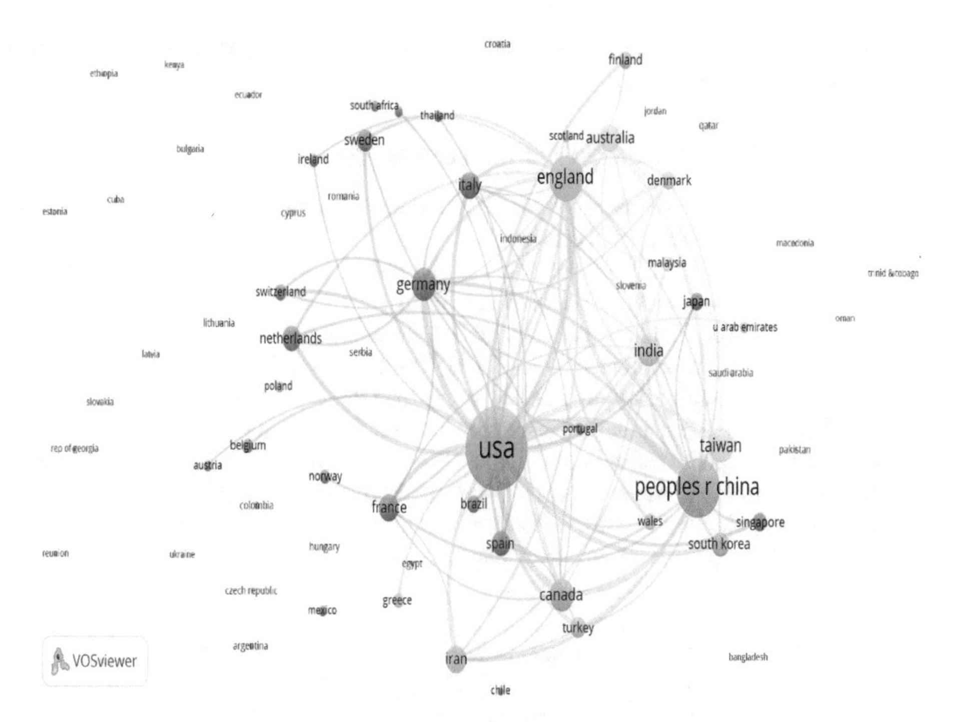

Fig. 1. Bibliographic coupling of the countries in SCM research- threshold: 96, connections: 100

on financial, societal, environmental etc. intentions. In this work during three periods from "2010 to 2017" this analysis is done. Respectively, Table 2 gives a comprehensive report about these three periods.

The USA as global leader country in each of these three periods also is in the first position among others but the other countries have experienced some changes in terms of ranking during this almost 30 years. As an example, China during the first period of 10 years has occupied the 15th position but this same country during the second period has achieved to the third position. Consecutively, according the last period after the USA, China is the second most productive country. The story for Taiwan is the same and period by period and year by year, this country has improved his position. Besides, UK as a developed country during these three periods has kept his position always among top countries of each period. Two other good examples that are suitable to be analyzed are India and Iran. Whilst during the period of 1990 and 1999, these two countries don't have significant positions, but with a fast looking the next periods, their growth in SCM area seem prominent.

Same as a global analysis, the figures of bibliographic coupling for countries is done for each of the periods to give a deeper and clearer insight. As an example, between "2010 to 2017" China and Taiwan have experienced a significant growth in comparison with the previous period. Iran and India are two samples of this newly comer countries on top positions of ranking. Figure 2 reports "2010–2017" with the threshold numbers 88 and the max 100 connections.

3.3 Individual Journal Analysis of the Leading Countries

The last section of the analysis is dedicated to the leading countries and their positions in some prominent journals of the area.

Table 3 reports the results of the two very important journals which dedicate their mission to publish as much as possible papers in the SCM research are. These two journals are "Supply Chain Management: An International Journal (SCMIJ)" and "Journal of Supply Chain Management (JSCM)". In SCMIJ, UK is the most productive country in terms of number of publications better than the USA. This point is important while, although the population of the USA is much more than UK, but the USA is the second country in this list. But in JSCM the USA has published 5 times more than the second country that is Germany and 10 times bigger than UK in terms of total number of publications. Another significant information from this column is the only 25 countries that have published at least 1 paper in this journal.

Table 2. Leading countries in SCM research between 2010 and 2017

R	Country	TPSCM	TCSCM	HSCM
1	USA	4287	64,512	93
2	China	2689	32,587	70
3	UK	1628	24,222	64
4	Taiwan	873	10,864	45
5	India	842	7,968	39
6	Germany	811	12,192	50
7	Canada	741	11,108	44
8	Iran	673	7,992	41
9	Australia	590	7,067	40
10	Italy	574	7,676	41
11	France	561	6,445	38
12	Spain	524	6,179	33
13	Netherlands	448	5,828	35
14	South Korea	381	3,500	28
15	Sweden	379	4,629	31
16	Turkey	323	3,921	31
17	Brazil	293	2,534	24
18	Denmark	286	5,727	36
19	Finland	237	2,397	24
20	Singapore	219	2,581	26

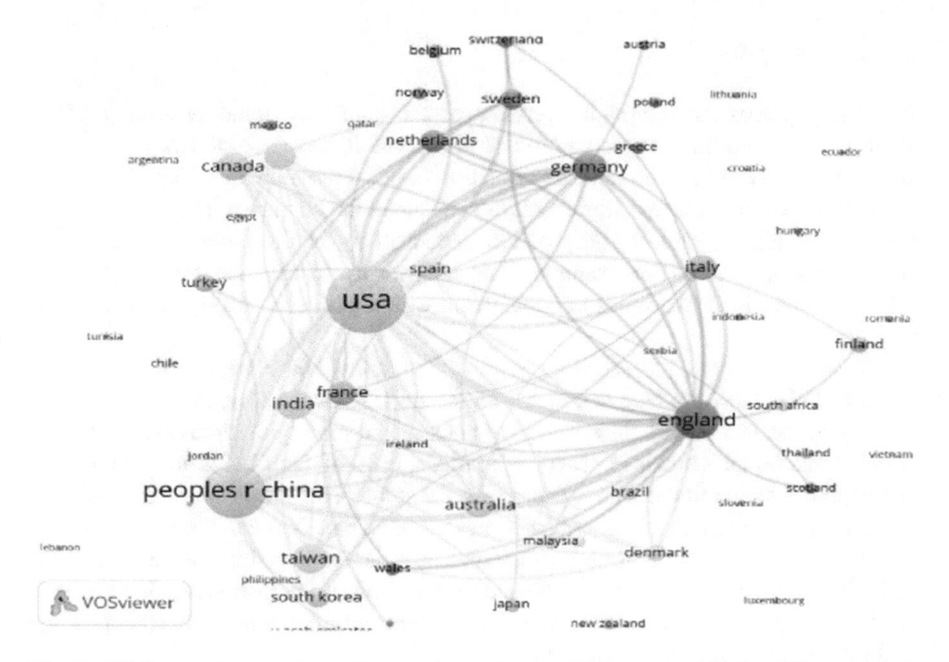

Fig. 2. Bibliographic coupling of countries between 2010 and 2017 in SCM research- *threshold: 88, connections: 100*

Table 3. Leading countries in SCM in supply chain management: an international journal and journal of supply chain management

Supply Chain Management an International Journal				Journal of Supply Chain Management		
R	Country	TP	TC	Country	TP	TC
1	UK	212	5,071	USA	156	5,396
2	USA	144	5,139	Germany	30	606
3	Australia	53	1,840	UK	16	443
4	China	44	1,099	Canada	15	937
5	Spain	42	805	China	10	226
6	Sweden	33	668	Switzerland	8	134
7	Taiwan	31	692	Netherlands	6	123
8	Netherlands	29	656	Spain	6	97
9	Finland	28	362	Australia	5	95
10	Germany	26	740	Ireland	5	181
11	Italy	24	722	South Korea	4	106
12	Canada	21	564	Denmark	3	43
13	Ireland	17	241	India	3	27
14	Denmark	14	471	Sweden	3	16
15	Norway	14	370	Brazil	2	20
16	Brazil	13	141	Finland	2	40
17	South Korea	13	755	Belgium	1	29
18	India	12	266	Chile	1	1
19	Greece	10	288	Ecuador	1	6
20	Malaysia	10	155	France	1	15

TP, TC: Total publications, total citations in SCM research.

4 Conclusions

This study presents a thorough overview of the countries' trend in Supply Chain Management discipline between 1990 and 2017. The first part of the analysis belongs to a global overview of the countries' situation in SCM research area. The idea behind this analysis is obtaining a comprehensive view about the countries. The USA not only globally but also in almost every period and in many of the individual journals is pioneer among others. For example, it is noteworthy to say that this country has published more than two times in comparison with China as a second country in this category.

Besides that, until the second period of the study UK always had the second position of the most productive countries after the USA when China has occupied his position in the last period. This trend is almost equal among the individual journals. From 14 individual journals or groups of journals, in 10 of them UK is among top three or sometimes as the first country. Moreover, for some other English-speaker countries like Canada, a certain trend cannot be expressed. Because, for example the position of this country throughout the time has experienced a descending slope however in some individual journals his position is still acceptable. One of the interesting points that is necessary to mention is an elevated level of connectivity between these countries that could be resulted from VOS viewer analysis.

Undoubtedly the most interesting point of this study is the enormous growth that the Asian countries have experienced. Among them China is the first country that from the 15th position in the period between 1990 and 1999, has achieved to the 2nd position in the last period between 2010 and 2017. Among the individual journals also China has a prominent position and almost in all of them is among 5 top productive countries. The other important Asian case is Taiwan that has experienced the similar growth like China during the time and in a global classification is the 4th country of the list. Among individual journals also, in many of them his position is acceptable. There are some other interesting countries like Iran and India that are among the countries that have experienced a significant growth in terms of total number of publications.

Although European countries in total have an acceptable portion from global point of view among others, but in comparison with Asian countries this portion is not that high. Besides UK, the most productive European country is Germany that has obtained the 6th position in global overview and after that, Italy, France and Netherlands. Although some Eastern European countries are among the top 50 countries in a global list, but their positions are not that high. However, the scenario is completely different for Scandinavian countries. They have gained better positions in comparison with Eastern European countries.

The situation of Latin American and African countries is not appropriate not at all. The best Latin American country is Brazil in terms of number of publications. But according to the investments in this field and paying an attention to improve the academic research, it is expectable that soon this situation will improve. The same story occurs for South Africa as the best African country. Although in one group of individual journals has obtained the second position, but the total number of publications for this country is lower than even the average. To justify the reason of this happening, in these countries based on the weak economy these results are normal and to improve their situation they need to develop a lot to reach the standards of developed nations (Confraria and Godinho 2015; Toivanen and Ponomariov 2011).

There is not an exact forecast for the future of the countries and there are several factors that can affect future that the most important one is the economic situation and the level of investment in the academic investigations. But as an example, it is predictable that some Asian countries like China based on their trends during the last years will experience the better positions and improve their situations while the other developed countries are also trying to improve their positions.

Besides, there are some limitations that should be considered. First, the diversity of the nationality of the researches in every university can be high and therefore the obtained result can't refer exactly to the host university also these publications can't be counted as an output of the country of author's origin. This happens a lot in the universities of the USA, that a plenty of the authors are not from the USA. Next, although our work is done on WoS database, but there are some other papers that could be in other databases, so it is possible to miss some of them. But it is important to say that this database is enormous enough to ignore this problem. The last limitation relates to those articles that have been published in other languages that are not English, so they are not included in WoS (Collazo-Reyes 2014). As a result, they are not considered as the publications of the country and the reality of the published papers and the number of the papers that exist in WoS can be different. However, it is important to repeat again that this database is huge enough to be considered as a confident reference of the work.

References

Alfalla-Luque, R., Medina-Lopez, C.: Supply chain management: unheard of in the 1970s, core to today's company. Bus. Hist. **51**, 202–221 (2009)

Barman, S., Hanna, M.D., LaForge, R.L.: Perceived relevance and quality of POM journals: a decade later. J. Oper. Manage. **19**, 367–385 (2001)

Broadus, R.: Toward a definition of "bibliometrics". Scientometrics **12**, 373–379 (1987)

Cancino, C., Merigó, J.M., Coronado, F., Dessouky, Y., Dessouky, M.: Forty years of computers & industrial engineering: a bibliometric analysis. Comput. Ind. Eng. **113**, 614–629 (2017)

Charvet, F.F., Cooper, M.C., Gardner, J.T.: The intellectual structure of the supply chain management: a bibliometric approach. J. Bus. Logistics **29**, 43–47 (2008)

Collazo-Reyes, F.: Growth of the number of indexed journals of Latin America and the Caribbean: the effect on the impact of each country. Scientometrics **98**, 197–209 (2014)

Confraria, H., Godinho, M.M.: The impact of African science: a bibliometric analysis. Scientometrics **102**, 1241–1268 (2015)

Cousins, P.D., Lawson, B., Squire, B.: Supply chain management: theory and practice–the emergence of an academic discipline? Int. J. Oper. Prod. Manage. **26**, 697–702 (2006)

Fry, T.D., Donohue, J.M.: Outlets for operations management research: a DEA assessment of journal quality and rankings. Int. J. Prod. Res. **51**, 7501–7526 (2013)

Goh, C.H., Holsapple, C.W., Johnson, L.E., Tanner, J.R.: Evaluating and classifying POM journals. J. Oper. Manage. **15**, 123–138 (1997)

Hirsch, J.E.: An index to quantify an individual's scientific research output. Proc. Nat. Acad. Sci. U.S.A. **102**, 16569 (2005)
Holsapple, C.W., Lee-Post, A.: Behavior-based analysis of knowledge dissemination channels in operations management. Omega **38**, 167–178 (2010)
Hsieh, P.N., Chang, P.L.: An assessment of world-wide research productivity in production and operations management. Int. J. Prod. Econ. **120**, 540–551 (2009)
Laengle, S., Merigó, J.M., Miranda, J., Słowiński, R., Bomze, I., Borgonovo, E., Dyson, R.G., Oliveira, J.F., Teunter, R.: Forty years of the European journal of operational research: a bibliometric overview. Eur. J. Oper. Res. **262**, 803–816 (2017)
Linderman, K., Chandrasekaran, A.: The scholarly exchange of knowledge in operations management. J. Oper. Manage. **28**, 357–366 (2010)
Martínez-López, F.J., Merigó, J.M., Valenzuela-Fernández, L., Nicolás, C.: Fifty years of the European journal of marketing: a bibliometric analysis. Eur. J. Mark. **52**, 439–468 (2018)
Merigó, J.M., Mas-Tur, A., Roig-Tierno, N., Ribeiro-Soriano, D.: A bibliometric overview of the journal of business research between 1973 and 2014. J. Bus. Res. **68**, 2645–2653 (2015)
Merigó, J.M., Yang, J.B.: A bibliometric analysis of operations research and management science. Omega **73**, 37–48 (2017)
Merigó, J.M., Pedrycz, W., Weber, R., de la Sotta, C.: Fifty years of information sciences: a bibliometric overview. Inf. Sci. **432**, 245–268 (2018)
Merino, M.T.G., Do Carmo, M.L.P., Álvarez, M.V.S.: 25 years of technovation: characterisation and evolution of the journal. Technovation **26**, 1303–1316 (2006)
Oliver, R.K., Webber, M.D.: Supply-chain management: logistics catches up with strategy. Outlook **5**, 42–47 (1982)
Petersen, C.G., Aase, G.R., Heiser, D.R.: Journal ranking analyses of operations management research. Int. J. Oper. Prod. Manage. **31**, 405–422 (2011)
Pilkington, A., Meredith, J.: The evolution of the intellectual structure of operations management – 1980–2006: A citation/co-citation analysis. J. Oper. Manage. **27**, 185–202 (2009)
Pritchard, A.: Statistical bibliography or bibliometrics. J. Documentation **25**, 348–349 (1969)
Shiau, W.L., Dwivedi, Y.K., Tsai, C.H.: Supply chain management: exploring the intellectual structure. Scientometrics **105**, 215–230 (2015)
Smith, J.S., Fox, G.L., Sunny Park, S.H., Lee, L.: Institutional antecedents to research productivity in operations: the US perspective. Int. J. Oper. Prod. Manage. **28**, 7–26 (2008)
Stonebraker, J.S., Gil, E., Kirkwood, C.W., Handfield, R.B.: Impact factor as a metric to assess journals where OM research is published. J. Oper. Manage. **30**, 24–43 (2012)
Theoharakis, V., Voss, C., Hadjinicola, G.C., Soteriou, A.C.: Insights into factors affecting production and operations management (POM) journal evaluation. J. Oper. Manage. **25**, 932–955 (2007)
Toivanen, H., Ponomariov, B.: African regional innovation systems: bibliometric analysis of research collaboration patterns 2005–2009. Scientometrics **88**, 471–493 (2011)
Van Eck, N.J., Waltman, L.: Software survey: VOS viewer, a computer program for bibliometric mapping. Scientometrics **84**, 523–538 (2010)

STEM Education: A Bibliometric Overview

Dolors Gil-Doménech[1(✉)], Jasmina Berbegal-Mirabent[1], and José M. Merigó[2,3]

[1] Department of Economy and Business Organisation, Universitat Internacional de Catalunya, C. Immaculada 22, 08017 Barcelona, Spain
{mdgil, jberbegal}@uic.es

[2] Department of Management Control and Information Systems, University of Chile, Av. Diagonal Paraguay 257, 8330015 Santiago, Chile
jmerigo@fen.uchile.cl

[3] School of Information, Systems and Modelling, Faculty of Engineering and Information Technology, University of Technology Sydney, 81 Broadway, Ultimo, NSW 2007, Australia

Abstract. In the recent years, STEM (science, technology, engineering and mathematics) education has received increasing attention, and many calls for a fundamental change in this field have emerged. In this context, the analysis of the evolution and development of this scientific domain results crucial. In order to do so, the present study presents a bibliometric overview of the academic research developed in STEM education over the last years. The work uses the Web of Science database and a wide range of bibliometric indicators including the number of publications and citations, the h-index, and citation thresholds. The article also develops a graphical visualisation of the bibliographic data using the visualisation of similarities (VOS) viewer software. Results indicate that the amount of works addressing this topic has substantially increased in the recent years, although the number of citations has not experienced a similar growth rate. Also, when analysing journals and research areas, it can be deduced that the research on STEM education is diverse. The bibliometric analysis performed provides a rigorous and comprehensive view of research on STEM education that might be useful for those researchers interested in advancing future knowledge in this area.

Keywords: STEM education · Bibliometrics · Journal analysis · Web of Science

1 Introduction

The term STEM arose in common use in the last 90s as an acronym for science, technology, engineering, and mathematics (Bybee 2010). As it involves all these disciplines, designing a curriculum or making decisions to promote STEM education implies a real challenge for educators and policymakers (Assefa and Abebe 2013).

In the recent years, the importance given to STEM education has experienced a significant growth, resulting in a number of calls for a fundamental change in the teaching of such disciplines (Eisenhart et al. 2015; Henderson et al. 2011). Given the

J. C. Ferrer-Comalat et al. (Eds.): MS-18 2018, AISC 894, pp. 193–205, 2020.
https://doi.org/10.1007/978-3-030-15413-4_15

increased attention this field has received—materialising in the design and implementation of new programs as well as of specific training for teaching STEM—, understanding the evolution and development of this scientific domain becomes critical. In this context, bibliometrics emerges as a science that permits understanding the main topics of a given domain, and how these topics relate to one another (Waltman et al. 2010).

Despite the existence of many studies providing general overviews on different aspects of STEM education (e.g. Holmegaard et al. 2014; Nugent et al. 2015), only a few of them analyse the state of art from a bibliometric perspective. On the one hand, there is a first bulk of studies that partially address this issue by using some elements of bibliometric analysis to understand the evolution of STEM education research. Specifically, Greenseid and Lawrenz (2011) aimed at assessing the influence of STEM education evaluations by using citation analysis methods. The authors concluded that the type of product (in their case, STEM education) as well as the program evaluation were significant predictors of the extent to which a product was cited. More recently, Assefa and Abebe (2013) performed a bibliometric mapping, using co-word analysis, of the structure of STEM education, and inferred that the core knowledge areas emerging from this domain include curriculum and professional development, policy formation, and resource management. On the other hand, other authors did apply bibliometric analysis in science-related education, but not explicitly on STEM. This is the case of Borrego and Bernhard (2011). By using bibliometrics they analysed engineering education research. The work of Barrow et al. (2008) also belongs to this set of studies. Concretely, these authors performed a bibliometric analysis to identify the major science education programs in the United States. Similarly, Dehdarirad et al. (2015), applied bibliometric tools to investigate on the role of women in science and higher education.

While the aforementioned papers offer a partial view of research on STEM education, this study aims at providing a comprehensive perspective. To do so, a general overview of STEM education using bibliometric analysis is presented, from its inception in 1991 until 2016. The underlying rationale is to identify the most productive and influential research in STEM education according to the information gathered from the Web of Science (WoS). This way, it will be possible to know the current evolution of this field.

Bibliometric analysis is an attractive research field for the scientific community. It allows for the identification, classification and analysis of bibliography, permitting the generation of summaries of the most outstanding results (Merigó and Yang 2017). Bibliometrics can be applied to any research field. As such, this methodology has been used in management (Podsakoff et al. 2008), entrepreneurship (Landström et al. 2012), or health economics (Wagstaff and Culyer 2012), among many others.

The rest of the chapter is organised as follows. In the next sections we present the results, including the most influential journals, the most cited papers, the most productive and influential authors and the main institutions in this area of knowledge. Next, we use the VOS viewer to graphically illustrate the bibliographic material. The chapter ends with the concluding remarks alongside with indicators for future works.

2 Bibliometrics and Citation Analysis

Since the strategy followed should cover all works related to STEM education, for the search process the keywords used were (educat* OR learn* OR teach*) crossed with (science* AND technolog* AND engineer* AND mathematic*). The search was conducted in January 2017, including all articles and reviews appearing in Web of Science (WoS) from its inception to December 2016. This search strategy resulted in 753 publications. Next, in an individual fashion, the authors of this work double-checked that the records obtained fit with the topic under analysis. To do this the reading of the title and the abstract was informative enough for making a decision. Only six articles did not match the topic, thus, the final sample comprises 747 articles and reviews.

As shown in Fig. 1, the majority of the works on STEM education have been published after 2010. More specifically, between years 2010 and 2016 there were 634 publications, representing 85% of the total volume, being 2016 the year with more publications (171, representing 23% of the total). Although it is obvious that this increase is mainly due to a growing interest in STEM education, it is also significant considering that both the number of researchers worldwide and the WoS itself have experienced an expansion during the last years, which have undoubtedly influenced the increase in the number of publications. By observing the exponential growth undergone in STEM education publications, it can be deduced that this field enjoys of good health inside the academic community.

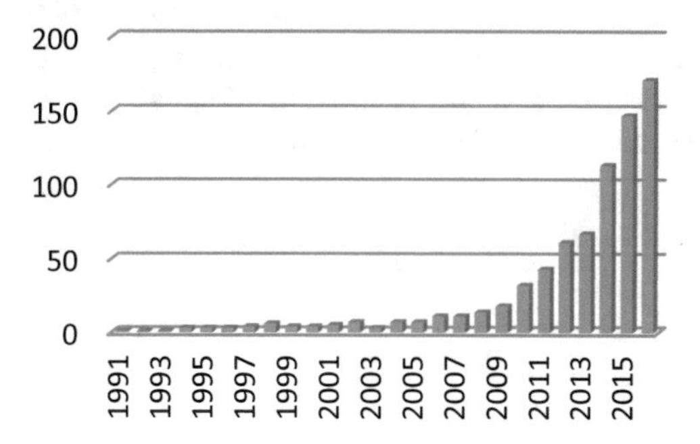

Fig. 1. Distribution of publications in STEM education (articles + reviews) per year, until 2016 in WoS

In order to evaluate the citation rate, Table 1 represents the general citation structure of all the papers on STEM education. As shown in the table, papers have been structured by thresholds concerning the number of citations. The percentage of papers in each category is also indicated.

Table 1. General citation structure in STEM education in WoS

Number of citations	Number of papers	% Papers
≥ 100 citations	8	1.071%
≥ 50 citations	20	2.677%
≥ 25 citations	47	6.292%
≥ 10 citations	117	15.663%
≥ 5 citations	199	26.640%
<5 citations	548	73.360%
Total	747	

Most of the papers (73.36%) received less than 5 citations. There are two potential explanations for this amount being that high. First, citations are not just a matter of quality but also of time. As papers on STEM are relatively new, the chances of being cited increases with time. Second, research on STEM is very broad, encompassing a variety of disciplines, ranging from education to computer since, but also including sociology and public administration, among other. The low citation rate thus seems to signal that works tend to be pretty specific. With respect to the most cited papers, it is remarkable that only eight papers received 100 citations or more, what represents 1.07% of the total sample, and only twenty (2.68%) received 50 citations or more.

3 Most Influential Journals in STEM Education

In order to identify the most influential journals in the field of study, Table 2 presents a list of the 15 journals with the highest number of papers in STEM education. The table includes different information of these journals. Journals are ranked based on the number of papers in the field of STEM education.

Table 2. 15 most influential journals in STEM education*

	Journal	Tot. Art.	Tot. Cit.	H-index
1	CBE-Life Sciences Education	43	144	7
2	J. of Science Education and Technology	31	65	5
3	Int. J. of Science Education	20	113	6
4	Int. J. of Engineering Education	19	49	4
5	J. of Engineering Education	17	309	11
6	Science Education	16	267	8
7	Int. J. of Technology and Design Education	16	45	4
8	IEEE Transactions on Education	13	108	5
9	J. of Chemical Education	11	40	4
10	Acta Astronautica	11	16	2
11	Theory Into Practice	10	13	2
12	J. of Educational Psychology	9	337	5
13	Computers & Education	9	24	3
14	Eurasia J. of Mathematics Science Tech Educ	9	11	2
15	J. for Multicultural Education	9	0	0

*The ranking is developed according to the total STEM education papers. In the case of a tie, the total STEM education citations are considered.

Considering the number of papers published in STEM education, *CBE-Life Sciences Education* and the *Journal of Science Education and Technology* seem to be the most influential journals in this field. This result is consistent with those presented in Table 1, as both journals are the outlet for two of the most cited papers in this field (with over 100 citations). Other relevant journals are the *Proceedings of the National Academy of Sciences of the United States of America*, the *Journal of Educational Psychology*, and the *Journal of Engineering Education*. This last journal, together with *Science Education* and the *International Journal of Science Education*, are the ones presenting a higher H-index in STEM education. Finally, it is remarkable that the journal with more influential papers in this field—see the next section below—is the *American Educational Research Journal*.

4 Most Influential Articles in STEM Education

The next step in this bibliometric analysis consists of identifying the most influential papers published in STEM education. To do so, a list with the 15 most cited papers of all time is presented in Table 3. Although there are many aspects that can influence the value of a paper, the number of citations is usually considered as a good reflection of the popularity and influence of a work among the scientific community (Merigó et al. 2015).

As it can be observed, the 1999 paper by Springer, Stanne, and Dodovan stands as the most cited work of all times, with 410 citations. Next in the raking we find the work of Freeman et al., published in 2014, with 251 citations. If we consider the rate citations per year, this work dominates this list, receiving an average of 84 citations per year.

5 Most Productive and Influential Authors in STEM Education

Another relevant issue in bibliometric analyses is to determine the most influential authors in the field. Table 4 shows the 15 authors with more papers in STEM education.

Note that Lou, with 6 papers, heads the list, followed by Smith, Tai, Tseng, MM Capraro, Shih, and RM Capraro, with 5 papers each. Also, it is remarkable that when the number of citations in STEM education is taken into account, the authors who stand out are Lubinski, Smith, and Uttal. All these authors should then be taken into account when listing the most influential authors in the STEM education field.

Table 3. 15 most cited papers in STEM education of all time

	Author/s	Journal	Tot. Cit.	Year
1	Springer, L; Stanne, ME; Donovan, SS	RER	410	1999
2	Freeman, S; Eddy, SL; McDonough, M; Smith, MK; Okoroafor, N; Jordt, H; Wenderoth, MP	PNAS	251	2014
3	Wai, J; Lubinski, D; Benbow, CP	JEP	224	2009
4	Ceci, SJ; Williams, WM	PNAS	172	2011
5	Blickenstaff, JC	GE	164	2005
6	Uttal, DH; Meadow, NG; Tipton, E; Hand, LL; Alden, AR; Warren, C; Newcombe, NS	PB	155	2013
7	Miyake, A; Kost-Smith, LE; Finkelstein, ND; Pollock, SJ; Cohen, GL; Ito, TA	Sci	133	2010
8	Brophy, S; Klein, S; Portsmore, M; Rogers, C	JEE	104	2008
9	Diekman, AB; Brown, ER; Johnston, AM; Clark, EK	PS	96	2010
10	Henderson, C; Beach, A; Finkelstein, N	JRST	93	2011
11	Reyna, VF; Brainerd, CJ	LID	90	2007
12	Maltese, AV; Tai, RH	SE	81	2011
13	Haak, DC; HilleRisLambers, J; Pitre, E; Freeman, S	Sci	75	2011
14	Harackiewicz, JM; Rozek, CS; Hulleman, CS; Hyde, JS	PS	70	2012
15	Zeldin, AL; Britner, SL; Pajares, F	JRST	67	2008

Abbreviations: RER, Review of Educational Research; PNAS, Proceedings of the National Academy of Sciences of the United States of America; JEP, Journal of Educational Psychology; GE, Gender and Education; PB, Psychological Bulletin; Sci, Science; JEE, Journal of Engineering Education; PS, Psychological Science; JRST, Journal of Research in Science Teaching; LID, Learning and Individual Differences; SE, Science Education.

Table 4. 15 most productive and influential authors in STEM education

	Author	Country	University	Tot. Art.
1	Lou SJ	Taiwan	National Pingtung University of Science and Technology	6
2	Smith MK	USA	University of Maine	5
3	Tai RH	USA	University of Virginia	5
4	Tseng KH	Taiwan	Meiho University	5
5	Capraro MM	USA	Texas A&M University	5
6	Shih RC	Taiwan	National Pingtung University of Science and Technology	5
7	Capraro RM	USA	Texas A&M University	5
8	Lubinski D	USA	Vanderbilt University	4
9	Uttal DH	USA	Northwestern University	4
10	Froyd JE	USA	Texas A&M University	4
11	Archer L	UK	King's College London	4
12	Dewitt J	UK	King's College London	4
13	Hurtado S	USA	University of California Los Angeles	4
14	Dillon J	UK	University of Bristol	4
15	Lopatto D	USA	Grinnell College	4

6 Most Productive and Influential Institutions in STEM Education

Next, Table 5 presents the list of the 15 institutions with more papers in STEM education. It is worth noting that all of these institutions are from the United States of America. The Purdue University, with 25 papers, heads the list of the most influential institutions in STEM education, followed by the University of Wisconsin, and the Texas A&M University, with 17 papers each. When it comes to consider the number of citations in this field, the University of Washington (366 citations), and the Northwestern University (334 citations) lead the classification.

Table 5. Ranking of the Top 15 Institutions

	University	Country	Tot. Art.	Tot. Cit.	H-index
1	Purdue Univ	USA	25	322	8
2	Univ Wisconsin	USA	17	218	8
3	Texas A&M Univ	USA	17	128	6
4	Northwestern Univ	USA	14	334	9
5	Univ Colorado	USA	14	292	5
6	Univ Virginia	USA	14	182	5
7	Univ Calif Santa Barbara	USA	14	42	4
8	Arizona State Univ	USA	13	40	3
9	Stanford Univ	USA	11	311	7
10	Univ Texas Austin	USA	11	106	5
11	Natl Sci Found (NSF)	USA	11	48	3
12	Univ Nebraska	USA	11	38	4
13	Michigan State Univ	USA	11	28	3
14	Univ Washington	USA	10	366	5
15	Univ Calif Los Ageles	USA	10	125	4

Crossing information from Tables 4 and 5, we observe that there is a correspondence between most prolific authors and universities. Specifically, three of the leading researchers in STEM education (MM Capraro, RM Capraro, and JE Froyd) are affiliated to University of Texas A&M (3rd in the ranking). Similarly, the National Pingtung University of Science and Technology is the home of SJ Lou, the researcher leading the list in Table 4, and RC Shih. The same applies to King's College London, with two outstanding authors in the field of STEM education (L Archer, and J Dewitt).

7 Analysis by Research Area in STEM Education

It is also relevant the analysis of the articles based on the research area. This approach allows for the identification of the main disciplines under which research in STEM education has been conducted. Accordingly, Table 7 presents the 15 most usual research areas where papers in STEM education can be classified according to the WoS.

Table 7. Most usual research areas (according to WoS) in STEM education papers

	Area	Tot. Art.	Tot. Cit.	H-index
1	Education Educational Research	454	3038	27
2	Engineering	138	742	15
3	Psychology	102	1334	17
4	Computer Science	60	146	7
5	Science Technology Other Topics	34	778	8
6	Social Sciences Other Topics	21	54	5
7	Information Science Library Science	14	53	4
8	Business Economics	13	88	4
9	Chemistry	13	43	4
10	History Philosophy of Science	13	10	2
11	Sociology	11	71	4
12	Mathematics	11	32	2
13	Life Sciences Biomedicine Other Topics	11	21	3
14	Materials Science	9	6	2
15	Environmental Sciences Ecology	8	14	2

As it can be seen, "Education and Educational Research" is, by far, the most usual research area, with 454 articles in STEM education classified in this area. This result is not surprising as the topic under analysis has an important pedagogical component. Another research area that outstands is "Psychology", with 102 terms. This field is close to education, therefore, both areas are concerned with the study of the human behaviour, particularly shaping the learning processes of students. As expected, another relevant research area is "Engineering", with 138 papers. Indeed, "Engineering" is one of the four fields that conforms the acronym of STEM.

8 Mapping STEM Education Research with VOS Viewer Software

In order to deepen into the results of the previous sections, this section graphically illustrates and analyses the bibliographic data on STEM education. To do so, the VOS viewer software is used (van Eck and Waltman 2010) which, based on all the documents that conform the sample, analyses co-citation, bibliographic coupling and co-occurrence of author keywords.

The co-citation analysis can be conducted among journals (Tang and Tsai 2016). Figure 2 shows the results considering a threshold of ten citations and the one hundred most representative connections.

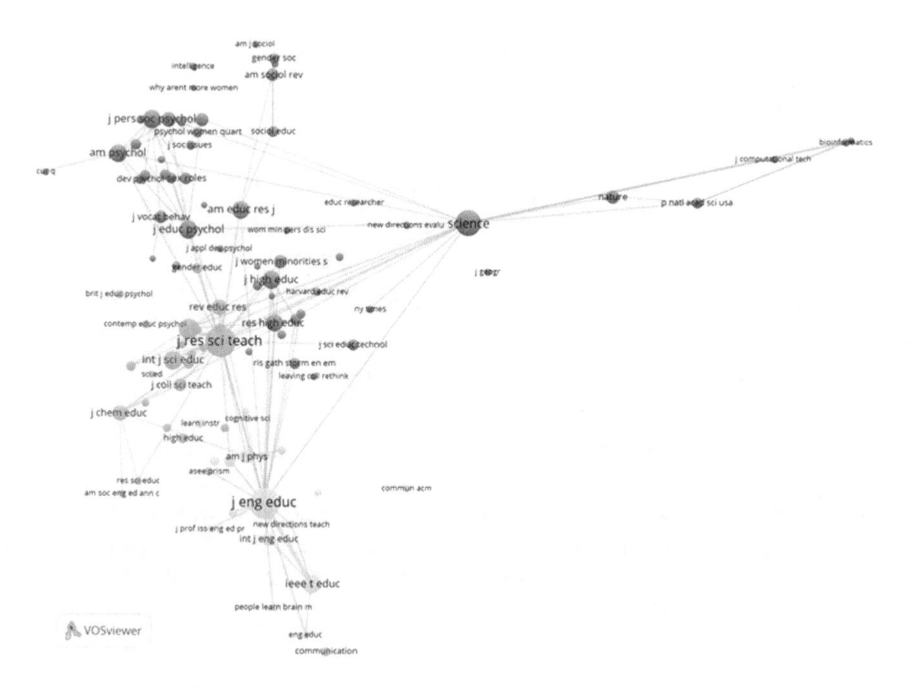

Fig. 2. Co-citation of journals among STEM publications

Journals in science and engineering education are the most cited journals with the strongest network connections. With respect to psychological journals, they are placed in the top-left of the figure. As found when analysing the research areas, research on psychology also plays an important role in STEM education.

It is also worth examining how the most productive institutions are connected to each other. For doing so, bibliographic coupling of institutions is used (Kessler 1963). Thus, Fig. 3 shows the results with a threshold of two documents and one hundred connections.

USA institutions act as major players, with very few universities outside of the USA appearing in Fig. 3. Among USA universities, it is worth noting Purdue University, University of Wisconsin, University of Virginia and Stanford University. Among the non-American institutions, University of Oslo (Norway), King's College London (United Kingdom) and Queensland University of Technology (Australia) can be found.

VOS viewer also allows for the analysis of the most frequent keywords appearing in a given set of publications. For performing this analysis, VOS viewer considers the keywords that appear either in the abstract, the first page of the article or the keywords

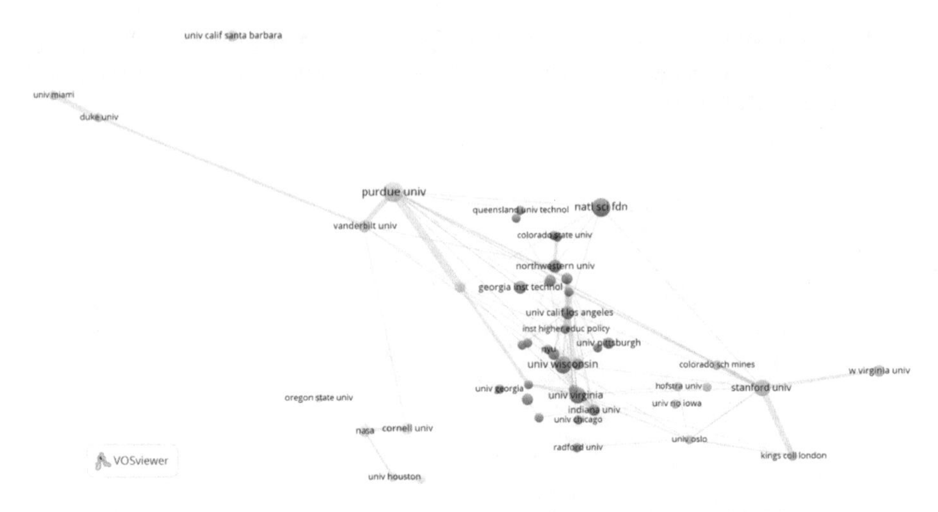

Fig. 3. Bibliographic coupling of the most productive institutions in STEM education

used by WoS in order to classify all the papers indexed in their database (Keyword Plus). For the purpose of this work, the focus is placed on keywords given by the authors in the title page, as this approach avoids the automatic selection of frequent words that appear in abstract but that do not provide insightful information about the real content of the publication. Figure 4 presents the results with a threshold of two documents and one hundred connections.

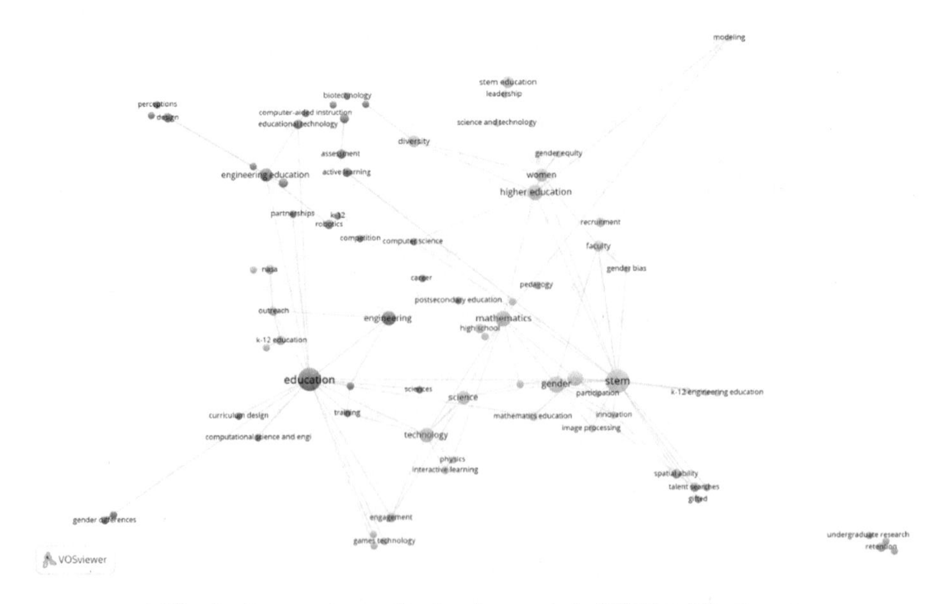

Fig. 4. Co-occurrence of author keywords in STEM publications

As shown in Fig. 4, “STEM” and “education” are the most frequent keywords. Some other popular keywords in terms of co-occurrence are “mathematics”, “engineering”, “higher education” and “engineering education”. This confirms that research on STEM education has an interdisciplinary perspective, as it connects with a wide range of fields including biotechnology, physics and psychology, among others.

9 Conclusions

This study offers an overview of research on STEM education using bibliometric techniques. From the analysis conducted several implications can be drawn. First, in the amount of works addressing this topic has substantially increased in the recent years. Nevertheless, it is worth highlighting that the number of citations that these papers have received has not experienced a similar growth rate. Accordingly, there is a limited number of papers with more than 100 citations.

The typical outlet for research on STEM education is diverse. Results indicate that there is no specific journal that outperforms the others. This way, we observe that the top 10 journals only concentrate 26% of the total production on STEM education. This diversity is also mirrored in the research areas. Specifically results reveal that STEM education embraces a wide variety of fields, ranging from education to business economics, among others.

The supremacy of USA is found at the institution level, almost having a monopoly. All the 15 most influential institutions are located in USA, being the Purdue University, the University of Wisconsin and the Texas A&M University leading the list. USA most representative authors are M.K. Smith and R.H. Tai from University of Maine and University of Virginia, respectively. Researchers from Texas A&M University (3rd in the ranking of most influential institutions) appear 5th (M.M. Capraro) and 7th in the list (R.M. Capraro). With respect to authors from other countries, we can find three of the most prolific authors from UK (L. Archer, J. Dewitt and J. Dillon) and other three from Taiwan (S.J. Lou, 1st in the list, K. H. Tseng and R. C. Shih).

Bibliometric studies permit obtaining a general picture of the state of the art of a given area or topic, nevertheless, several limitations due to the research strategy followed and how documents have been classified should not be ignored. In this respect, it is important noting that there are many other databases that could have been used for this study, such as Scopus or Google Scholar. By doing so, relevant works published in books or presented at academic conferences are underestimated. Likewise, contributions written in a different language than English are not considered. Due to the relevance of WoS and following the current standards of how research is generally assessed, we decided to focus on this database, and more specifically to its Core Collection. Second, from a general perspective, research on STEM education is a very broad area that includes a wide range of topics. Due to this breadth, it is important not to underestimate the relevance of works that, for instance, because of the specific topic, have received less citations that another one. Third, when analysing the most influential researchers, our analysis does not take into account the number of researchers that contributed in writing a specific publication. Therefore, single-authored papers are equally accounted as multiple-authored works. This problem is also present when analysing results by country. WoS gives one unit to each participating country of an article, regardless the number of countries participating in the research study. This fact might bring some deviations.

Finally, it is worth noting that quantifying research is not an easy task. Thus, despite the aforementioned limitations, we believe this article provides a sound and comprehensive view of research on STEM education that might be helpful for those researchers interested in advancing future knowledge in this area. In future research, we expect to develop further developments to this approach by analysing other particular cases and separating the four key categories that constitute the Business & Economics area. Moreover, we will also consider other countries and journals in the analysis and extend this approach to other scientific categories.

References

Assefa, S.G., Abebe, R.: A bibliometric mapping of the structure of STEM education using co-word analysis. J. Am. Soc. Inf. Sci. Technol. **64**(12), 2513–2536 (2013)

Barrow, L.H., Settlage, J., Germann, P.J.: Institutional research productivity in science education for the 1990s: top 30 rankings. J. Sci. Educ. Technol. **17**, 357–365 (2008)

Borrego, M., Bernhard, J.: The emergence of engineering education research as an internationally connected field of inquiry. J. Eng. Educ. **100**(1), 14–47 (2011)

Bybee, R.W.: Advancing STEM education: a 2020 vision. Technol. Eng. Teach. **70**(1), 30–35 (2010)

Dehdarirad, T., Villarroya, A., Barrios, M.: Research on women in science and higher education: a bibliometric analysis. Scientometrics **103**, 795–812 (2015)

Eisenhart, M., Weis, L., Allen, C.D., Cipollone, K., Stich, A., Dominguez, R.: High school opportunities for STEM: Comparing inclusive STEM-focused and comprehensive high schools in two US cities. J. Res. Sci. Teach. **52**(6), 763–789 (2015)

Greenseid, L.O., Lawrenz, F.: Using citation analysis methods to assess the influence of science, technology, engineering, and mathematics education evaluations. Am. J. Eval. **32**(3), 392–407 (2011)

Henderson, C., Beach, A., Finkelstein, N.: Facilitating change in undergraduate STEM instructional practices: an analytic review of the literature. J. Res. Sci. Teach. **48**(8), 952–984 (2011)

Holmegaard, H.T., Madsen, L.M., Ulriksen, L.: To choose or not to choose science: constructions of desirable identities among young people considering a STEM higher education programme. Int. J. Sci. Educ. **36**(2), 186–215 (2014)

Kessler, M.M.: Bibliographic coupling between scientific papers. Am. Documentation **14**(1), 10–25 (1963)

Landström, H., Harirchi, G., Åström, F.: Entrepreneurship: exploring the knowledge base. Res. Policy **41**(7), 1154–1181 (2012)

Merigó, J.M., Gil-Lafuente, A.M., Yager, R.R.: An overview of fuzzy research with bibliometric indicators. Appl. Soft Comput. **27**, 420–433 (2015)

Merigó, J.M., Yang, J.-B.: A bibliometric analysis of operations research and management science. Omega **73**, 37–48 (2017)

Nugent, G., Barker, B., Welch, G., Grandgenett, N., Wu, C., Nelson, C.: A model of factors contributing to STEM learning and career orientation. Int. J. Sci. Educ. **37**(7), 1067–1088 (2015)

Podsakoff, P.M., MacKenzie, S.B., Podsakoff, N.P., Bachrach, D.G.: Scholarly influence in the field of management: a bibliometric analysis of the determinants of university and author impact in the past quarter century. J. Manag. **3**(4), 641–720 (2008)

Tang, K.-Y., Tsai, C.-C.: The intellectual structure of research on Educational Technology in Science Education (ETiSE): a co-citation network analysis of publications in selected journals (2008–2013). J. Sci. Educ. Technol. **25**(2), 327–344 (2016)

van Eck, N.J., Waltman, L.: Software survey: VOSviewer, a computer program for bibliometric mapping. Scientometrics **84**(2), 523–538 (2010)

Wagstaff, A., Culyer, A.J.: Four decades of health economics through a bibliometric lens. J. Health Econ. **31**(2), 406–439 (2012)

Waltman, L., van Eck, N.J., Noyons, E.C.M.: A unified approach to mapping and clustering of bibliometric networks. J. Informetr. **4**(4), 629–635 (2010)

Premises for the Theory of Forgotten Effects

Salvador Linares-Mustarós[1(✉)], Anna M. Gil-Lafuente[2], Dolors Corominas Coll[1], and Joan C. Ferrer-Comalat[1]

[1] Department of Business Administration, University of Girona, Carrer de la Universitat de Girona 10, 17071 Girona, Spain
{salvador.linares,dolors.corominas, joancarles.ferrer}@udg.edu

[2] Department of Business Administration, University of Barcelona, Av. Diagonal 690, 08034 Barcelona, Spain
amgil@ub.edu

Abstract. This paper aims to present the logical foundations in research into forgotten effects by showing an essential corollary to understanding forgotten effects models. The corollary states that any influence that may occur between two events is really of a value greater than or equal to the maximum truth value of any chain of cause and effect between the two events. This is consistent with the fact that in practical research on forgotten effects, experts always change the value of an impact if they are shown a causal chain of more value than they have estimated directly.

Keywords: Forgotten effects · Computer software · Incidence matrices

1 Introduction

The idea of using fuzzy incidence composition to detect the possible non-obvious influences of a cause on an effect is known in the economic literature as "Forgotten Effects Research". The application of incidence matrices in an economic context was first presented to the scientific community in 1988 by Professors Arnold Kaufmann and Jaime Gil Aluja in the book published by Milladoiro "Models to research forgotten effects" (Kaufmann and Gil Aluja 1988).

There are a great number of academic papers (García de Fanelli 2001; Eugenio Mallo et al. 2005; Domech More et al. 2008; Gil Lafuente et al. 2009; Rico et al. 2010; Gil Lafuente and Luis Bassa 2011; Gil Lafuente et al. 2012; Linares et al. 2013) and research projects (Lladó et al. 2000; Gil Lafuente et al. 2011, 2012) addressing indirect incidence values based on various models presented by researchers.

Our aim in this paper consists in presenting the logical foundations in research into forgotten effects. To achieve our objective, the chapter has been divided into two sections. The first section presents the terminology used throughout this work and an overview of how the method usually functions. The second section presents the basis of binary logic that supports the theory of research into forgotten effects. The paper ends with conclusions and the bibliography.

J. C. Ferrer-Comalat et al. (Eds.): MS-18 2018, AISC 894, pp. 206–215, 2020.
https://doi.org/10.1007/978-3-030-15413-4_16

2 Presentation of the Usual Method Employed to Search for Forgotten Effects Based on Estimates Provided by Experts

Analysis of the concept of fuzzy relationships was first undertaken by Zadeh in the 1971 article “Similarity relations and fuzzy orderings” (Zadeh 1971), extending ideas conceived a few years previously in the seminal article “Fuzzy logic” (Zadeh 1965).

In 1988, Arnold Kaufmann and Jaime Gil Aluja established (Kaufmann and Gil 1988) the “Theory of Forgotten Effects” using models which allow all incidence relationships or direct and indirect causalities to be obtained without the possibility of omitting elements that might initially have been fully or partially overlooked.

The following summarizes the general fuzzy rectangular matrix model that will be used to model the specific problem in the following section.

Let A be a set of elements, $A = \{a_i / i = 1, 2, \ldots, n\}$, which we shall call causes, and B, a second set of elements, $B = \{b_j / j = 1, 2, \ldots, m\}$, which could possibly be the same set as A, and we shall call effects.

The resulting set of pairs of elements denoted by the expression $v(a_i, b_j)$ shall define what tends to be known as the “direct incidence matrix”. This matrix shows the cause-effect relationships that occur between the elements of set A (causes) and those in set B (effects). Said matrix will be denoted by $[\underline{M}]$.

Now let C be a third set of elements, $C = \{c_k / k = 1, 2, \ldots, p\}$, which act as effects of set B (their acting as causes of set A would be treated similarly).

Let $[\underline{N}]$ be the incidence matrix that shows the cause-effect relationships occurring between the elements of set B and those of set C.

The mathematical operator that establishes the incidences of A on C is the max-min composition of matrices, whose causal relationship of a_i over c_k is:

$$v(a_i, c_k) = \max_{b_j}\left(\min\left(v\left(a_i, b_j\right), v\left(b_j, c_k\right)\right)\right)$$

Causes-causes will be noted in the incidence matrix for the whole as $[\underline{A}]$ and effects-effects as $[\underline{B}]$. The composition $[\underline{M}^*] = [\underline{A}] \circ [\underline{M}] \circ [\underline{B}]$ between these three types of matrices is designed to obtain a new matrix of incidences from among the elements of set A (causes) and those of set B (effects) to reflect any possible indirect causal relationships that may not be taken into account in strategic decision processes.

To find possible effects which may have been overlooked, the algorithm proposed by Kaufmann and Gil suggests studying the values of the matrix $[\underline{M}^*] - [\underline{M}]$, as the values obtained thus are maximum values of indirect effects.

The fact that values close to one appear in the matrix $[\underline{M}^*] - [\underline{M}]$ indicates that the initial incidents have not been correctly estimated as there are some omissions.

Gento et al. (2001) pointed out that when creating a direct incidence matrix thanks to the method of recovering forgotten effects, the construction of a new direct incidence matrix results in greater internal coherence with respect to the basis for recovering intermediate effects, obtaining as a result “few forgotten effects”.

Given that in problems involving the selection of certain actions that produce certain benefits for a greater number of actions, overlooked incidents can lead to choosing a less efficient set of actions, the aforementioned fact indicates that research into forgotten effects is an essential method for improving the information the decision-maker has to work with.

In studies carried out in the economic sphere, different values tend to be used for incidences of the same cause of the same effect due to ignorance of actual incidence values. These values are provided by a group of experts and aggregated together using special expert management techniques, in order to obtain a single value for each incidence. One of the most common techniques for adding information together in such works is the expertage technique, first proposed by Kaufmann (1987) and developed theoretically by the same author and Professor Gil Aluja (Kaufmann and Gil Aluja 1993).

In most work on forgotten effects it is also common for experts' valuation of incidence to be a value belonging to the interval [0, 1]. Due to this, researchers tend to provide experts with a linguistic table with a numerical scale similar to Table 1, known as a linguistic endecadary scale. The model of this scale was presented by Kaufmann and Gil Aluja in the book "Introduction to the Theory of Fuzzy Subsets in Business Management" (Kaufmann and Gil Aluja, 1986).

Table 1. Version of the classic endecadary scale

Null incidence	0
Practically null incidence	0.1
Almost null incidence	0.2
Very weak incidence	0.3
Weak incidence	0.4
Intermediate incidence	0.5
Fair incidence	0.6
Considerable incidence	0.7
Strong incidence	0.8
Very strong incidence	0.9
Absolute incidence	1

There is also a second endecadary scale to determine whether the values experts assign to incidence are correct. The main difference between this scale, presented at the 16th INTERNATIONAL SIGEF CONGRESS (Bonet et al. 2010), and Kaufmann and Gil Aluja's classic scale, is how the former determines subjective inputs. This is achieved via a maximum two-stage process using a ternary selection system after eliminating extreme cases. In the first phase of allocation, there is reflection on the degree of truth inherent in the statement, choosing from the possibilities low (L), medium (M) and high (H), once the possibilities of completely false (N for null) and completely true (T for total) have been rejected. If situated within L, M or H, in a second phase, the question is asked whether comparing with the other reference elements believed to be in the same group, the element being evaluated is within the subgroup low (l), medium (m) or high (h).

The final allocation is obtained from the Table 2 of values.

Table 2. Version of the double ternary scale

N	L			M			H			T
	l	m	h	l	m	h	l	m	h	
0	0.1	0.2	0.3	0.4	0.5	0.6	0.7	0.8	0.9	1

Thus, for example, if we aim to associate the possibility of whether incidence that improved water treatment in a certain country might lead to the effect of improving health, we can, in the cases where null (N) and all (T) are rejected, as a first step choose between low (L), medium (M) or high (H). Let us imagine that we have chosen high (H). Then we enter a second phase considering whether this high will be very (h), a little (l) or neither one nor the other (m).

We can establish a bijective application h: S → T between the two scales that retains the following order property:

$$h(u) \leq_T h(v) \text{ if and only if } u \leq_S v$$

It can be concluded that an order isomorphism exists between the two endecadary scales, presented in Table 3.

Table 3. Order isomorphism

(S, ≤S)		(T, ≤T)
Null incidence	0	N
Practically null incidence	0.1	Ll
Almost null incidence	0.2	Lm
Incidence quite close to null	0.3	Lh
Incidence closer to null than full	0.4	Ml
Neither null nor full incidence	0.5	Mm
Incidence closer to full than null	0.6	Mh
Incidence quite close to full	0.7	Hl
Almost full incidence	0.8	Hm
Practically full incidence	0.9	Hh
Full incidence	1	T

3 A Proposal for Basing the Theory on Sentential Logic

Definition 1. A declarative sentence is a sentence in the form of a statement. Consequently, a sentence is declarative if and only if it only affirms or only denies some objectively verifiable fact. Therefore, in the same context, the truth value of these statements never admits various interpretations since it is well established. The truth value of a declarative sentence "A" is denoted by "v(A)" and this may be "T" or "1" if the declarative sentence is true or "⊥" or "0" if it is false.

For example, the legal definition of entering adulthood usually varies between ages 16–21, depending on the region. If we set a city and a time, "Person X is an adult" is a declarative sentence.

Note 1. "It's raining" or "It's raining outside" are two declarative sentences that have the same meaning. In this case, we talk about declarative sentences or propositions being equivalent. To simplify the calculations, we assume that when a set of declarative sentences is formed, there are no two equivalent statements.

Definition 2. A declarative sentence is composed if and only if it can be obtained by connecting one or two declarative sentences with one of the following particles: "not", "and", "or" or "if … then …".

Note 2. The connector "not" is called negation.

Note 3. The connector "and" is called conjunction.

Note 4. The connector "or" is called (inclusive) disjunction.

Definition 3. The connector "if … then …" is called conditional. And in this sense, in the expression "if A, then B", "A" is termed the antecedent and "B" the consequent of the conditional.

Definition. Classical Propositional Calculus. To formalize the set of composed declarative sentences, the following assumptions are made.

- All sentences formed from the logical connective "not" applied to a declarative sentence are declarative sentences that have the following truth values:

A		not A (¬A)
1		0
0		1

- This table is read, for example, as "If the declarative sentence "A" is true, then the sentence "not A" is a false declarative sentence".
- All sentences formed from the logical connectors "… and …" "… or …" and "if … then …" applied to two declarative sentences are declarative sentences whose truth value is one of the following tables:

A	B		A and B ($A \wedge B$)
1	1		1
1	0		0
0	1		0
0	0		0

A	B		A or B ($A \vee B$)
1	1		1
1	0		1
0	1		1
0	0		0

A	B		If A, then B ($A \rightarrow B$)
1	1		1
1	0		0
0	1		1
0	0		1

Note 5: Pla (1991) showed that the structure of declarative sentences with connections "not" and "and" lets build "if … then …" connection and other new connections, as "or. .. or … " and " … if and only if … ". For example, the "if … then …" connection can be obtained with the following formative construction:

A	B		$\neg A \vee B$
1	1		1
1	0		0
0	1		1
0	0		1

Note 6: The fact of working with 0 and 1 instead of ⊥ and T allows the possibility of defining the evaluation of declarative sentences connection in terms of nonlinear mathematical functions. For example, the mathematical function of the ¬A operation can be defined as $v(\neg A) = 1 - v(A)$; $A \wedge B$ can be defined as

$v(A \wedge B) = \min\{v(A), v(B)\}$; $A \vee B$ can be defined as
$v(A \vee B) = \max\{v(A), v(B)\}$; and finally, $A \rightarrow B$ can be defined as
$v(A \rightarrow B) = \max\{1 - v(A), v(B)\}$.

Proposition 1. ∧ is associative.

Proof: It's trivial according to this composition table.

A	B	C		A∧B	(A∧B)∧C	B∧C	A∧(B∧C)
1	1	1		1	1	1	1
1	0	1		0	0	0	0
0	1	1		0	0	1	0
0	0	1		0	0	0	0
1	1	0		1	0	0	0
1	0	0		0	0	0	0
0	1	0		0	0	0	0
0	0	0		0	0	0	0

Proposition 2: If A, B and C are three declarative sentences, $(A \rightarrow B) \wedge (B \rightarrow C)$ is a declarative sentence.

Proof: $(A \rightarrow B) \wedge (B \rightarrow C)$ is a sentence composed of the sentences $A \rightarrow B$ and $B \rightarrow C$.
We know from Classical Propositional Calculus that $A \rightarrow B$ and $B \rightarrow C$ are declarative sentences. Consequently, from Classical Propositional Calculus, their connection with ∧ is also a declarative sentence.

Note 4: The next proposition allows the modelling of Hume's idea about the connection between two objects related to a third object, admitted by Hume as a principle of human nature. Specifically, Hume wrote "…we must consider, that two objects are connected together in the imagination, not only when the one is immediately resembling, contiguous to, or the cause of the other, but also when there is interposed betwixt them a third object, which bears to both of them any of these relations" (Hume 1739).

Proposition 3: If A, B and C are three declarative sentences, then
$v[A \rightarrow C] \geq v[(A \rightarrow B) \wedge (B \rightarrow C)]$.

Proof: It's trivial according to this composition table.

A	B	C		A→B	B→C	(A→B)∧(B→C)	A→C
1	1	1		1	1	1	1
1	0	1		0	1	0	1
0	1	1		1	1	1	1
0	0	1		1	1	1	1
1	1	0		1	0	0	0
1	0	0		0	1	0	0
0	1	0		1	0	0	1
0	0	0		1	1	1	1

Proposition 4: If A_1, A_2, …, A_n are declarative sentences, then $(A_1 \to A_2) \wedge (A_2 \to A_3) \wedge \ldots \wedge (A_{n-2} \to A_{n-1}) \wedge (A_{n-1} \to A_n)$ is a declarative sentence.

Proof: by mathematical induction.

Step 1. Show it is true for n = 2.
The second proposition shows it is true.
Step 2. Assume it is true for $n - 1 \geq 2$.
Then $(A_1 \to A_2) \wedge (A_2 \to A_3) \wedge \ldots \wedge (A_{n-2} \to A_{n-1})$ is a declarative sentence.
$A_{n-1} \to A_n$ is a declarative sentence according to Classical Propositional Calculus. Therefore, according to Classical Propositional Calculus the union (∧) between $(A_1 \to A_2) \wedge (A_2 \to A_3) \wedge \ldots \wedge (A_{n-2} \to A_{n-1})$ and $A_{n-1} \to A_n$ is a declarative sentence. So, $(A_1 \to A_2) \wedge (A_2 \to A_3) \wedge \ldots \wedge (A_{n-2} \to A_{n-1}) \wedge (A_{n-1} \to A_n)$ is a declarative sentence.

Proposition 5: If A_1, A_2, …, A_n are declarative sentences, then
$$v[A_1 \to A_n] \geq v[(A_1 \to A_2) \wedge (A_2 \to A_3) \wedge \ldots \wedge (A_{n-2} \to A_{n-1}) \wedge (A_{n-1} \to A_n)].$$

Proof: by mathematical induction.

Step 1. Show it is true for **n = 2**.
The third proposition shows it is true.
Step 2. Assume it is true for $n - 1 \geq 2$.

$$\text{Then } v[A_1 \to A_{n-1}] \geq v[(A_1 \to A_2) \wedge (A_2 \to A_3) \wedge \ldots \wedge (A_{n-2} \to A_{n-1})] = \max[v[A_1 \to A_2], \ldots, v[A_{n-1} \to A_n]],$$

and $v[A_1 \to A_n] \geq v[(A_1 \to A_{n-1}) \wedge (A_{n-1} \to A_n)] = \max[v[A_1 \to A_{n-1}], v[A_{n-1} \to A_n]]$, so,

$$\begin{aligned} v[A_1 \to A_n] &\geq \max[v[A_1 \to A_{n-1}], v[A_{n-1} \to A_n]] \\ &\geq \max[\max[v[A_1 \to A_2], \ldots, v[A_{n-1} \to A_n]], v[A_{n-1} \to A_n]] \\ &= \max[v[A_1 \to A_2], \ldots, v[A_{n-1} \to A_n], v[A_{n-1} \to A_n]] \\ &= v[(A_1 \to A_2) \wedge (A_2 \to A_3) \wedge \ldots \wedge (A_{n-2} \to A_{n-1})]. \end{aligned}$$

Corollary: If A_1, A_2, $A'_2, \ldots, A_{n-1}$, A'_{n-1}, A_n are declarative sentences, then

$$\begin{aligned} v[A_1 \rightarrow A_n] \geq\ & \max(v[(A_1 \rightarrow A_2) \wedge (A_2 \rightarrow A_3) \wedge \ldots \wedge (A_{n-2} \rightarrow A_{n-1}) \wedge (A_{n-1} \rightarrow A_n)], \\ & v[(A_1 \rightarrow A'_2) \wedge (A'_2 \rightarrow A'_3) \wedge \ldots \wedge (A'_{n-2} \rightarrow A'_{n-1}) \wedge (A'_{n-1} \rightarrow A_n)]) \\ =\ & \max(\min[v(A_1 \rightarrow A_2), v(A_2 \rightarrow A_3), \ldots, v(A_{n-2} \rightarrow A_{n-1}), v\ (A_{n-1} \rightarrow A_n)], \\ & \min[v(A_1 \rightarrow A'_2), v(A'_2 \rightarrow A'_3), \ldots, v(A'_{n-2} \rightarrow A'_{n-1}), v(A'_{n-1} \rightarrow A_n)]) \end{aligned}$$

4 Conclusion

This paper has shown that in binary logic any influence that may occur between two events is a truth value greater or equal to the maximum truth value of any chain of cause and effect between the two given events. This idea is essential to understanding the theory of forgotten effects. To test whether this applies to fuzzy logic is the future goal of this study.

References

Anselin Ávila, E., Gil Lafuente, A.M.: Fuzzy logic in the strategic analysis: impact of the external factors over business. Int. J. Bus. Innov. Res. **5**(3), 515–534 (2009)

Bonet, J., Linares, S., Bertran, X.: Cálculo mixto estadístico-fuzzy, por comparación, aplicado a la previsión de variables económicas. In: Proceedings of the XVI International Congress of the SIGEF on Economical and Financial Systems in Emergent Economies, Morelia Michoacán, Mexico (2010)

Domech More, J., Carvalho Aguia, J.: Modelo Fuzzy de Influência entre Indicadores de Responsabilidade Social, Simpósio de Excelência em Gestão e Tecnologia (SEGeT), Rio de Janeiro, Brasil (2008)

Eugenio Mallo, P., Antonia Artola, M., Javier Galante, M., Enrique Pascual, M., Morettini, M., Raúl Busetto, A.: Aplicación de herramientas borrosas al balance scorecard. In: XXVIII congreso argentino de profesores universitarios de costos. Mendoza, Argentina (2005)

García, P.S., Lazarri, L.L., Machado, E.A.M.: Una propuesta fuzzy para definir indicadores de pobreza. Cuadernos de CIMBAGE **3**, 11–26 (2000)

García, P.S., Lazarri, L.L.: La evaluación de la calidad en la Universidad. Cuadernos de CIMBAGE **3**, 81–97 (2000)

García de Fanelli, A.M.: Aplicación de la metodología borrosa a la evaluación de la calidad de los posgrados. Cuadernos de CIMBAGE **4**, 99–132 (2001)

Gento, A., Lazzari, L.L., Machado, E.A.M.: Reflexiones acerca de las matrices de incidencia y la recuperación de efectos olvidados. Cuadernos de CIMBAGE **4**, 11–27 (2001)

Gil Lafuente, A.M., Vizuete Luciano, E., Martorell Cunill, O.: Modelo para la determinación de variables que inciden en el consumo de productos financieros de ahorro. In: XV SIGEF International Conference, Lugo, Spain (2009)

Gil Lafuente, A.M., Luis Bassa, C.: Proceso de identificación de los atributos contemplados por los clientes en una estrategia CRM. Cuadernos de CIMBAGE **13**, 107–127 (2011)

Gil Lafuente, A.M., Amiguet Molina, L., Vizuete Luciano, E., Boria Reverter, S., Sole Moro, M. L., Pursals Puig, A., Imanov, G., Aliyev, T., Rzayev Rza, R., Akbarov Mammad, R., Yusifzada Arshad, R.: Nuevos mercados para la recuperación económica: Azerbaiyán, Publicaciones de la Real Academia de Ciencias Económicas y Financieras, Barcelona (2011)

Gil Lafuente, A.M., Gil Lafuente, J., Vizuete Luciano, E., Amiguet Molina, L., Boria Reverter, S., Merigó Lindahl, J.M., Klimova, A., Sole Moro, M.L.: Explorando nuevos mercados: Ucrania, Publicaciones de la Real Academia de Ciencias Económicas y Financieras, Barcelona (2012)

Kaufmann, A.: Les expertons. Ed. Hermes, París (1987)

Kaufmann, A., Gil Aluja, J.: Introducción de la Teoría de los Subconjuntos Borrosos a la Gestión de las Empresas, Milladoiro, Santiago de Compostela (1986)

Kaufmann, A., Gil Aluja, J.: Modelos para la investigación de efectos olvidados, Milladoiro, Santiago de Compostela (1988)

Lazzari, L.L.: La asignación de recursos en acciones tendientes a mejorar la imagen de una PYME. In: VI Congreso de la Pequeña y Mediana Empresa, Buenos Aires, Argentina (2000)

Lladó, M., Gil Aluja, J., Rabaseda i Tarrés, J., Ferrer Comalat, J.C., Gil Lafuente, A.M., Gil Lafuente, J.: Determinació d'accions per a l'ascens de categoria del Girona FC i la seva reprecussió sobre l'àrea territorial Girona-Costa Brava. Publicacions Departament Empresa, Girona (2000)

Rodríguez Rubinos, J.M., Armando Ramírez Reyes, M., Díaz Pontones, V.: Efectos olvidados en las relaciones de causalidad de las acciones del sistema de capacitación en las organizaciones empresariales. Revista de Métodos Cuantitativos para la Economía y la Empresa **5**, 29–48 (2008)

Rico, F., Marco, A., Jaime Tinto, A.: Herramientas con base en subconjuntos borrosos. Propuesta procedimental para aplicar expertizaje y recuperar efectos olvidados en la información contable. Actualidad Contable FACES **13**(21), 127–146 (2010)

Zadeh, L.A.: Similarity relations and fuzzy orderings. Inf. Sci. **3**(2), 177–200 (1971)

Zadeh, L.A.: Fuzzy sets. Inf. Control **8**, 338–353 (1965)

Author Index

J. C. Ferrer-Comalat et al. (Eds.): MS-18 2018, AISC 894, pp. 217–218, 2020.
https://doi.org/10.1007/978-3-030-15413-4

Printed in the United States
By Bookmasters